SCRUM QUEST

NAVIGATING AGILE PRODUCT DELIVERY WITH CLARITY AND CONFIDENCE, ONE SPRINT AT A TIME

A HANDS-ON SIMULATION GUIDE FOR MASTERING SCRUM THROUGH REAL-WORLD JOURNEY

SCRUM QUEST

The Quick Start Edition

Written by Nan D. Ross

Printed Worldwide

First Printing 2026

First Edition 2026

10 9 8 7 6 5 4 3 2 1

Published by CORPORATE COSMO, LLC.

TABLE OF CONTENTS

1 INTRODUCTION:

YOUR AGILE MAP AND COMPASS

The room was alive with a mix of caffeine and friction. Team Vertex had gathered for Sprint Planning, laptops open, a half-empty coffee cup leaving rings on the table, sticky notes scattered across the table like crude waypoints, and the backlog projected on the screen. Maria, the Product Owner, was trying to prioritize features for the new mobile app, but the list kept growing.

Alex, the developer, was eager to dive in and start coding the login screen immediately. Jenna, the designer, argued that without a clear onboarding trail or flow before any code was written, users would feel lost. Malik, the QA engineer, raised concerns that none of the acceptance criteria were testable yet, warning that without checkpoints, the team would stumble later. Maya, the Scrum Master, tried to reset the compass and keep the discussion on course, but conversations circled back to the same debates.

What should have been a focused planning session felt instead like setting off into the wild without a map or markers. Deadlines loomed and confidence wavered. The team's excitement shifted toward frustration. It is evident that without clear direction, even small choices can become costly detours.

RAISE THE STAKES

Left unchecked, this isn't just a bad meeting; it's a pattern that eats delivery. Priorities shift, unfinished work piles up, and rework swallows momentum. Teams lose rhythm, stakeholders grow impatient, and the product's value gets delayed. Agile is not just a nice-to-have framework; it's the difference between an expedition that reaches the summit and one that returns empty-handed.

That's where *Scrum Quest* comes in. This book is not a dry manual or a wall of principles. It is your **field guide to the Agile Software Development Life Cycle (SDLC)**, presented as an expedition. You'll carry two essential tools on the journey:

- **Your map** – the Agile-aligned SDLC, broken into clear checkpoints: Planning, Requirements, Design, Development, Testing, Deployment, and Maintenance.
- **Your compass** – the Scrum framework - always points you back to values, principles, and collaboration when the path ahead looks unclear.

This isn't theory folded into platitudes. Instead of trudging through jargon-heavy lectures, you'll join **Team Vertex**, a fictional yet realistic cross-functional product team, as they chart their course one Sprint at a time. Through their journey, you'll encounter the obstacles and turning points of modern product development: missed deadlines, clashing perspectives, tough prioritization calls— moments that turn a team from stumbling hikers into a confident expedition.

YOUR CHALLENGE ALONG THE WAY

This isn't a story to skim, you've gotta live it. *Scrum Quest* asks you to step into the team's decisions and ask: *What choice would I make in their shoes?* When the backlog feels too big or priorities shift, which path do you pick? When QA finds late surprises, how do you respond?

Each chapter gives you the chance to test your instincts, compare them with the team's choices, and refine your own Agile approach.

YOUR EXPEDITION GEAR

No explorer travels light or sets out without the right tools, and neither will you. As part of this journey, you'll get access to the very same artifacts that guide Team Vertex on their quest to deliver their **next mobile app**. These are working deliverables you can study, adapt, and apply to your own practice:

- **Team Vertex Product Backlog** – a living map of features, epics, and priorities.
- **Sprint Backlogs & User Stories** – the checkpoints that keep the team moving forward.
- **Workflow Diagrams & Wireframes** – visual trail markers that bring clarity to process and design.
- **Live Prototype** – a working model you can explore and critique just as the team does.
- **Test Cases & Test Scripts** – the safety ropes ensuring quality at every step.
- **Code Review Examples** – a behind-the-scenes look at how the team inspects and adapts at the technical level.

These artifacts are your practical, hands-on expedition gear that turn theory into application and make the journey real.

FROM KNOWLEDGE TO TRANSFORMATION

This book is more than just a compendium of practices - it's your training ground. By the time you finish *Scrum Quest*, you won't simply understand Agile concepts; you'll have lived them through story, reflection, and practice. You'll move from uncertainty to confidence, from theory to application, and from being a participant in Agile delivery to a true guide.

Bonus: you'll gain an inside look at the **key practices of the Development Team,** seeing how the **designer, developer, and QA** each contribute in every Sprint. You'll observe how design shapes user experience, how code is written and reviewed, and how testing safeguards quality. More importantly, you'll understand how these roles collaborate, overlap, and lean on each other to keep the expedition moving forward.

Whether your goal is to step into a new Scrum role, sharpen your leadership, or elevate your team's delivery, this journey equips you with the tools, scenarios, and mindset shifts to level up professionally, transforming how you think, collaborate, and deliver value in the real world.

TRAIL MAP OF LEARNING: WHAT YOU'LL GAIN ON THE JOURNEY

Every expedition has checkpoints with moments to pause, regroup, and mark progress. In *Scrum Quest*, each Sprint is a checkpoint in your Agile journey. Here's what you'll discover as you travel with Team Vertex through the full Agile SDLC:

SPRINT ZERO: PREPARING FOR THE JOURNEY

Before setting out, the team equips itself with the essentials. You'll learn how to:

- Craft the Agile Project Scope to define purpose and scope.
- Map the terrain with stakeholders to align expectations.
- Estimate using story points and spikes.
- Set the compass with working agreements and the Definition of Ready/Done.
- Stock the backpacks by creating the initial backlog and capacity plans.
- Chart the trail with workflow diagrams for requirements.
- Slice epics into features and features into user stories.

OUTCOME: A shared sense of direction, confidence, and clarity before the first step.

SPRINT ONE: THE FIRST STEP ON THE TRAIL

The journey begins with real delivery. You'll explore how to:

- Select the right trailhead by delivering a small, high-value increment (Profile Creation & Login).
- Plan the expedition leg through Sprint Planning and backlog commitment.
- Unpack user stories and refine them for development and testing.
- Navigate daily checkpoints with Daily Scrums.
- Build, test, and inspect the first increment.
- Review the trail behind you in a Sprint Review.
- Rest and reflect at camp in a Retrospective.

OUTCOME: Momentum, alignment, and the first working product increment.

SPRINT TWO: TACKLING THE NEXT ASCENT

The climb gets steeper as the team builds **Log Feeding**. You'll learn how to:

- Refine the trail map by slicing feeding-related stories.
- Balance conflicting stakeholder voices.
- Pair design, development, and QA for incremental progress.
- Adapt the compass mid-route as backlogs evolve.
- Test in real terrain with integration and regression testing.
- Gather trail feedback through Sprint Reviews.
- Reset packs and prepare for the next ascent during Retrospectives.

OUTCOME: A more complex feature delivered while strengthening teamwork and adaptability.

SPRINT THREE: NAVIGATING ROCKY TERRAIN

The path turns rugged with **Walk Log** - timers, GPS, and user interactions. You'll uncover how to:

- Handle complex trails with dependencies and technical hurdles.
- Chart workflows visually to refine requirements.
- Use burn charts as trail markers for pacing.
- Collaborate under pressure when blockers arise.
- Incorporate user feedback to validate direction.
- Conduct User Acceptance Testing (UAT) to confirm business readiness before release.
- Reset the compass with advanced Retrospective formats.

OUTCOME: The ability to manage complexity, validate with UAT, and stay aligned under pressure.

SPRINT FOUR: REACHING NEW HEIGHTS

The final MVP push: **Log Medication**. Precision matters as the stakes rise. You'll learn how to:

- Slice work for maximum value while avoiding overload.
- Strengthen safety ropes with test cases, scripts, and reviews.
- Balance value with velocity to maximize delivery.
- Demonstrate product readiness at the Sprint Review.
- Elevate team growth through deeper Retrospectives.
- Deploy to Production (MVP release), delivering a real, usable product to end users.
- See the trail behind you by reflecting on the entire expedition.

OUTCOME: A complete, reliable MVP in production and a stronger, more confident Agile team.

BY THE END OF THE TRAIL

From Sprint Zero to Sprint Four, you'll have experienced every phase of the Agile software development lifecycle through Scrum. Each checkpoint builds not only the PetTracker app but also your mastery.

THE JOURNEY AHEAD

By the final chapter, you will understand Agile software development and know how to lead the quest. You'll be equipped to guide teams through uncertainty, align stakeholders around value, and deliver working solutions that matter. Whether you're a new Scrum Master, a Product Owner sharpening your practice, or a professional crossing into Agile delivery, *Scrum Quest* equips you with the confidence**, clarity, and compass you need to succeed one Sprint at a time.**

Pack your gear. Check your compass. Let the expedition begin!

2.
MEET TEAM VERTEX

Picture this: You walk into your first team meeting at BrightByte Solutions expecting some structure: maybe an agenda, a round of introductions, or at least someone acknowledging the new face in the room.

Instead, you walk straight into a fire.

The product demo is minutes away. The lead developer is rewriting code live. The designer is still scrambling with last-minute edits in Figma. QA is pointing out a bug, but no one has time to fix it. And worse, someone just asked, "Wait... what exactly are we showing today?"

Welcome to Team Vertex, a group full of talent, but without a consistent way to work. They care deeply about their craft, but their operation methods need a lot of work.

No one's clear on the goal. Deadlines slip. Priorities change mid-week. What gets built rarely matches what was originally asked for. People are burning out, and everyone's quietly wondering: **How long can we keep this up?**

THE REALITY CHECK

Best belief, they know something needs improvement, but they have already tried project plans, status meetings, and a dozen different tools. Still, they are misaligned and firefighting amidst the chaos of deadlines.

As another last-minute production change spins out of control, Maya, the new Scrum Master, remarks with calm frustration, "We can't keep working like this; I've seen teams turn this around with clarity and accountability." She blows out a breath before adding, "We need a structure that supports change, instead of breaking under it."

Her suggestion? **Scrum** as a practical framework to bring focus, rhythm, and sanity to the way Team Vertex works.

Some eyes roll. A few people sigh. The reactions are mixed. You recognize this situation: exhaustion from continuous effort has people weary of new proposals, regardless of merit or potential impact.

That's when Maya turns to you.

"You're new," she says. "You've got a fresh perspective. I want you to help us figure out how to introduce Scrum in a way that actually works for this team. Nothing performative. Focus on real results."

TIME TO STEP IN

As much as you are a part of the team, you're also a quiet agent of change.

Over the next few chapters, you'll get to know Team Vertex, explore what Scrum really is (and what it's not), and learn what it takes to turn good intentions into focused, sustainable momentum.

Along the way, you'll reflect on questions like:

- What would you do in this moment?
- How would you turn a skeptical team into believers of the values and principles of Agile?
- What necessary and honest conversations never happened?

MEET TEAM VERTEX: INSIDE THE MINDS BEHIND THE MISSION

"EVERY GREAT PRODUCT IS BUILT BY PEOPLE FIRST."

Before we jump into Sprint planning and stand-ups, let's meet the personalities behind the product and the tensions you'll help navigate.

Alex (The Tech Lead/Developer)

Alex carries the technical weight of the team on his back. He's sharp but burnt out from being everyone's "go-to." He loves clean code, hates chaos, and has little patience for vague requirements or surprise changes.

WHAT HE NEEDS: A trusted structure and a team that can deliver without using him as a permanent fixer.

Jenna (The UX Designer)

She's your creative and passionate team member. Jenna's designs are clean and intuitive, but they're often done last-minute because no one includes her early enough. She's tired of being treated like an afterthought.

WHAT SHE NEEDS: A seat at the table from day one and a process that values user experience, not just features.

Malik (The QA Analyst)

Malik has a sixth sense for catching bugs, but he's often handed builds too late to test properly. He's always the last to touch things and somehow the first blamed when things break.

WHAT HE NEEDS: Time to test, visibility into what's coming, and to stop being treated as the team's safety net.

Maria (The Product Owner)

She carries pressure like her shadow. Maria tries to be an efficient bridge between shareholders and the team. But priorities shift fast, and she's often changing direction faster than the team can keep up.

WHAT SHE NEEDS: A stable process that lets her manage change without derailing everything else and a better way to translate stories the team can build on with confidence.

Maya (The Scrum Master)

Maya doesn't bark orders, yet her influence is felt. She's the team's facilitator, coach, and protector of process. While others chase delivery, she's making sure they don't burn out or break down.

WHAT SHE NEEDS: A team that embraces continuous improvement and the space to remove real blockers, not just perform rituals.

You (The New Team Member)

As Maya said, you're the fresh perspective. You're joining as the newest team member and bringing ideas, asking questions, and helping guide the team's next steps.

WHAT YOU NEED: A clear framework, a collaborative environment, and a real opportunity to help build something meaningful without the usual noise.

Together, this crew makes up **Team Vertex**, a group with the potential to do great things... if they can just get out of their own way and harmonize.

WHAT'S NEXT?

As Team Vertex wraps up their last Sprint, they realize they've mastered the mechanics but missed the heartbeat. In the next chapter, we'll explore the values agile teams live by, revealing how trust, collaboration, and continuous feedback transform rituals into results and why embracing these values is the true catalyst for high-performing teams.

Let's dig in.

3. THE VALUES AGILE TEAMS LIVE BY

Mindset matters just as much as the tools used in soft delivery. Sadly, teams often prioritize tools over mindset. Team Vertex had all the rituals in place but lacked the underlying principles to turn process into progress.

In this chapter, we define the Software Development Life Cycle SDLC as the essential work required to move an idea into production work that exists in both Waterfall and Agile even when Agile teams use different names for the lifecycle. You will see how Agile does not eliminate these phases but approaches them iteratively using values and Scrum behaviors to reshape how teams collaborate adapt and deliver real customer value. When teams share these values alignment not rigid steps becomes the true catalyst for high performing delivery.

AGILE = A MINDSET, NOT JUST A PROCESS

First things first, let's define Agile in a sentence:

AGILE IS AN ITERATIVE WAY OF WORKING THAT HELPS TEAMS DELIVER VALUE IN SMALL INCREMENTS WHILE LEARNING AND ADAPTING THROUGH CONTINUOUS FEEDBACK.LY.

That mindset guides every practice, from backlog refinement to continuous integration. Teams that don't get Agile's "why" fall back on repeated ceremonies and lose sight of genuine collaboration.

BEFORE THE FRAMEWORK: THE AGILE MINDSET

Before the boards, stand-ups, and sticky notes, there's the matter of mindset. **Agile** is a value system that shapes how teams think, collaborate, and deliver. For Team Vertex, the early days were filled with motion but not momentum. They had tools and ceremonies, but lacked the foundation that makes Agile truly work: shared values.

THE FOUR FOUNDATIONAL VALUES (AGILE MANIFESTO)

1. **Individuals and interactions** over processes and tools
2. **Working software** over comprehensive documentation
3. **Customer collaboration** over contract negotiation
4. **Responding to change** over following a plan

Team Vertex lacked trust and a shared understanding. Agile flips that script.

Values in the Wild: How Team Vertex Was (and Wasn't) Living Them

In those first few weeks, Team Vertex had stand-ups on the calendar, Jira tickets in motion, and retros scheduled. On paper, it looked like progress. Under the surface, cracks were forming.

- **Individuals and interactions?**
  Slack threads had replaced real conversations. Maria (PO) felt disconnected from the devs. Jenna designed in isolation. "We're collaborating" meant handing things off to one another.
- **Working software?**
  Stories were "done," but features were only partially tested, and the login flow still broke on mobile. They were busy but not delivering value.
- **Customer collaboration?**
  No one had spoken to a user. Feedback came secondhand through support tickets or assumptions.
- **Responding to change?**
  Mid-Sprint changes sparked confusion. Stories ballooned. Instead of adapting, the team stalled, unsure how to shift without spiraling.

Maya noticed the tension. Instead of enforcing more processes, she got curious. Effort was not an issue; the problem was an absence of alignment.

SCRUM = HOW WE BEHAVE

Agile gives us the "why." Scrum gives us the "how." While Agile lays the foundation, Scrum builds the daily rhythm and behaviors that bring it to life. It requires teams to show up differently.

At the core of Scrum are five values that shape not just what the team does, but how they do it, especially when things get messy or unpredictable.

THE FIVE SCRUM VALUES

- **Commitment:** Showing up for the team and the goal
- **Focus:** Working toward one Sprint Goal at a time
- **Openness:** Sharing progress, blockers, and feedback honestly
- **Respect:** Valuing every role on the team
- **Courage:** Facing tough problems and having real conversations

Maya didn't name them yet, but she modeled them when she said, "This isn't working: let's try something different."

WHEN VALUES ARE MISSING, AND IT SHOWS

Before Team Vertex learned the language of Scrum, their behavior spoke in ways that didn't help them grow. Early struggles were value failures: misalignment, avoidance, and silence. These moments were signals.

TEAM VERTEX'S EARLY CONFLICTS TIED TO MISSING VALUES

Conflict	Missing Value(s)	What It Cost Them
Alex and Jenna clashed on handoff timing; designs came late.	Openness & Respect	Frustration, rework, breakdown in team trust.
Maria pushed new features mid-Sprint.	Focus & Commitment	Scope creep, burnout, missed Sprint Goals.
No one raised a late-stage blocker until the Sprint Review.	Courage	QA stress, failed demo, loss of confidence.
Maya ran ceremonies by the book, but no one engaged meaningfully.	Openness & Commitment	Going through motions without real improvement.
Malik found bugs but didn't speak up until the last day of testing.	Openness & Courage	Last-minute panic, reduced release quality.

Maya saw it happening. The structure was there. The meetings were booked. But the team wasn't connecting or growing. She shifted the focus to cultivating behavior and getting the team to live Scrum.

After the Workshop: Quiet Realization

The room had mostly cleared in Scrum. Whiteboard scribbles of "Working Software," "Individuals and Interactions," and "Responding to Change" still glowed faintly under the overhead lights.

Maya lingered, erasing the board slowly, deep in thought. She thought to herself: If we can actually live this, like really live it, we might finally break the cycle.

She glanced at a sticky note Maria had scribbled during the exercise:

'Trust > Tasks.'

Maya tucked it into her notebook.

The team wasn't there yet. But something had shifted today.

Not everyone saw it. But she did.

What's Next? Turning Realizations into a New Way of Working

That quiet but foundational shift is what makes Agile more than a buzzword. In the next chapter, we'll examine the "why" behind Agile's rise. Before Team Vertex could adopt Scrum with intention, they needed to confront what wasn't working in their current approach and why Agile offered more than just a new process. It offered a new way forward.

Exercise: Early Lessons of Values in Action

On a prior engagement, Team Vertex launched the **Workout Scheduler** module for the FitLife fitness app. Although it was declared "done," several breakdowns surfaced:

1. **Maria** skipped backlog refinement, assuming "schedule a workout" was self-explanatory.
2. **Jenna** designed the calendar interface alone, without consulting QA or Developers.
3. **Alex** coded the scheduling logic but omitted tests for time-zone edge cases.
4. **Malik** discovered the app's reminder-notification rules only after testing began.

Instructions

1. Identify which **Agile or Scrum Value(s)** were missing in each situation.
2. Explain how practicing that value could have improved the outcome.

Situation	Missing Value(s)	How It Could've Helped
Maria skipped refinement		
Jenna worked in isolation		
Alex omitted edge-case tests		
Malik learned the rules late		
No stakeholder demo held		

4.

WHY AGILE? DISCOVERING A NEW WAY

Before a team adopts Agile, the work feels like a treadmill: exhausting, repetitive, and oddly stationary. That's where Team Vertex found themselves. Despite talent and effort, their results lagged behind expectations. Deadlines loomed, quality suffered, and frustration mounted. But beneath the chaos was a system that couldn't flex when reality did. This chapter exposes the turning point where pain meets possibility, and a new way of working begins to take root. Agile wasn't a buzzword anymore. It was a necessary shift in how they thought, planned, and delivered.

TEAM VERTEX'S CHAOS AND FRUSTRATION

It's 9:03 AM.

The daily meeting was supposed to start at 9 sharp, but half the team is still hunting for coffee or rushing to patch bugs that should've been handled two weeks ago. The designer is frustrated, the developer is confused, and the project manager is exhausted. And no one can explain why the last release broke production again.

Team Vertex is doing what most teams do before they find Agile: working hard but running in circles. Requirements change mid-Sprint. Priorities shift without warning. "Is this even in scope?" has become the team's unofficial motto.

Everyone wants to build something great, but doesn't know how to build it together.

Sound familiar?

That's the moment when real change becomes possible because the pain finally pushes the team to say: **"There has to be a better way."**

AGILE VS. WATERFALL: MOMENT OF TRUTH

Before Agile, most teams used a method called **Waterfall**. Think of it like a relay race: one person runs their leg, hands the baton to the next, and the process continues linearly.

It looks like this:

1. **Plan everything up front.**
2. **Design it all.**
3. **Build it all.**
4. **Test at the end.**
5. **Release: ready or not.**

It sounds organized in theory. But here's the problem: **What if the customer changes their mind halfway through the race?**

Toss in a shrug that says, "Too bad, Buddy." Waterfall teams keep running the planned path even if the finish line has moved.

Agile does things differently. Instead of locking everything down, Agile embraces feedback, **welcomes change**, and **delivers in small, testable pieces**. So, if something shifts, teams can shift with it before it's too late.

Waterfall	Agile
Big upfront planning	Just enough to get started
Sequential handoffs	Cross-functional collaboration
Change avoided	Change welcomed
One big release	Continuous small releases

TRADITIONAL WATERFALL VS SCRUM

Where Team Vertex Got Stuck in Waterfall Thinking

Team Vertex didn't realize it, but they were *working like a Waterfall team with Agile labels.*

They planned everything at once, hoping to avoid change. Designers and Developers worked in isolation. Testing happened last minute. The backlog was treated like a static scope doc, not a living source of evolving needs. "Done" meant "developer finished," not "value delivered."

Maya later reflected on it: *"We were calling it a Sprint. But we were still building in a straight line without any real feedback loops."*

WHAT AGILE WOULD'VE LOOKED LIKE

If Team Vertex had been working in true Agile rhythm:

- Maria would've prioritized a few **high-value stories**, not handed over a 3-month roadmap.
- Jenna would've looped in Alex during wireframes, not after polish.
- Malik would've started testing flows as soon as they were buildable, not a day before release.
- They would've demoed working features early, gathered feedback, and iterated quickly.

Team Vertex needed flow, feedback, and focus.

THE PROFILE UPDATE FEATURE

Team Vertex gets a request from stakeholders: "Let users update their profile: name, email, photo, and preferences."

How They Handled It Before (Waterfall Thinking)

1. **Maria** (PO) writes a full spec doc with every requirement spelled out.
2. **Jenna** (Designer) disappears into Figma for 2 weeks and hands over polished screens with no team input.
3. **Alex** (Dev) gets the design, finds gaps, and makes assumptions. Backend APIs aren't ready.
4. **Malik** (QA) gets the build days before launch and finds major bugs, but it's too late to fix them cleanly.
5. **Maya** (SM) pushes to meet the deadline anyway. The feature ships were half-tested.
6. Users complain that photo upload fails on mobile, and preferences don't save.

It would eventually be one big handoff that failed woefully.

How They Could've Handled It (Agile Flow)

1. **Maria** shares the *user problem*, not a full-blown spec: *"Users want to update basic profile info to personalize their account."*
2. The team **discusses it together** in Backlog Refinement. They break it into three stories:
   - Update name + email
   - Upload/change profile photo
   - Adjust notification preferences
3. **Jenna** creates a quick wireframe. **Alex** reviews it early and flags API needs.
4. They build and test the *first story* in Sprint 1. **Malik** writes tests alongside development.

5. After the Sprint Review, users confirm: "This is what we need. Can we make the photo step mobile-first?"
6. The team adjusts before Sprint 2.

Everything happens in smaller chunks. There's clearer collaboration, and everyone has a smile on their faces.

Maya's Takeaway to the Team:

"You build trust with feedback loops that are fast, messy, and honest."

When Process Masks the Problem

Team Vertex didn't realize it yet, but they were **missing the mindset** entirely. Beneath the chaos, the symptoms were all there: reactive planning, buried blockers, misaligned handoffs. They were working hard, but rarely working together.

The issue was glaring, and the **Agile Principles** had already warned about what happens when teams ignore flow, feedback, and focus.

Let's look at the warning signs that Team Vertex was experiencing and what the Agile lens reveals beneath the surface.

Signs of Misalignment: What the Agile Principles Warned About

Agile requires a mindset shift. It's a response to common dysfunctions teams face when speed and pressure outweigh purpose and process. The Agile Manifesto includes 12 principles that call attention to what *should* be happening in healthy teams.

Team Vertex was nowhere near that. Here's what the newest team member noticed and what the Agile principles would've flagged right away.

---

1. Meetings with No Purpose

Daily standups had devolved into status monologues.

"Why are we meeting if no one's actually syncing?"

Agile Principle #4: *"Business people and developers must work together daily."*

---

2. Priorities Were a Moving Target

The backlog was overloaded, outdated, or both. Stories changed mid-Sprint.

"I just started this ticket yesterday. Now we're de-prioritizing it?"

Agile Principle #2: *"Welcome changing requirements, even late in development."*

---

3. Too Much Work in Progress

Everyone was juggling five things at once, yet nothing was finished.

"We've started 12 things... have we finished even one?"

Agile Principle #8: *"Agile processes promote sustainable development."*

---

4. Blockers Were Silent

People hit walls but didn't speak up.

"How long have we been stuck on this bug?"

Agile Principle #12: *"At regular intervals, the team reflects and adjusts."*

---

5. UX & Dev Handoff Chaos

Designs arrived half-explained late in the Sprint.

"Wait, where's the spec for this layout? Oh... there isn't one."

Agile Principle #6: *"Face-to-face conversation is the best form of communication."*

---

6. QA Was Always an Afterthought

Testing came late, and it was often rushed and reactive.

"We're releasing tomorrow, and QA just got the build?"

Agile Principle #7: *"Working software is the primary measure of progress."*

---

7. "Done" Meant Different Things

One dev shipped it. Another thought it needed QA. The PO thought it wasn't even ready.

"I thought this was finished?" "Nope, it hasn't been tested or reviewed."

Agile Principle #1: *"Our highest priority is to satisfy the customer through early and* *continuous delivery."*

This was the wake-up call. Scrum wasn't going to fix the team on its own.

They needed to address what they were doing *wrong* before any of the framework could start working *right*.

Team Vertex didn't need another all-hands meeting. They needed alignment with principles that could actually **change how they worked and how they worked together**.

Maya chipped in, "We're stuck because we keep solving symptoms instead of the system. Until we change how we *work together*, we'll keep Sprinting in circles."

WHAT'S NEXT: FROM MINDSET TO METHOD: ENTER SCRUM

Agile gave Team Vertex the *why*, but they needed a framework for *how*. That's where Scrum stepped in. It offered a lightweight, repeatable structure that transformed reactive chaos into purposeful rhythm. In the next chapter, Maya introduces the team to Scrum's roles, events, and artifacts, showing them how to move from theory to real traction, one Sprint at a time.

Exercise: Waterfall Thinking vs. Agile Flow: Can You Spot the Difference?

Team Vertex is about to tackle their **first feature: Profile Creation/Login**. You're given two approaches below. **Review each and answer the questions that follow.**

Scenario A: Waterfall Thinking

1. The Product Owner writes a full requirements document upfront.

2. The Designer creates finalized screens without team input.
3. Developers don't see the design until Sprint Day 4.
4. Testing starts the day before the Sprint ends.
5. Feedback is gathered after the release.

Scenario B: Agile Flow

1. The Product Owner shares the user need: "Users want to create an account and log in securely."
2. The team collaborates in Backlog Refinement to break it into smaller stories (e.g., Create Account, Login with Email, Error Handling).
3. Designers share early sketches and gather input from Dev and QA.
4. Developers build incrementally and testers validate early.
5. Feedback is gathered during the Sprint Review, and stories are improved for the next Sprint.

Reflection Questions:

1. Which scenario demonstrates Agile principles in action and why?
2. List at least two risks or bottlenecks in Scenario A that Agile helps prevent.
3. How does cross-functional teamwork in Scenario B support better Sprint outcomes?

5.

SCRUM 101

Before Team Vertex could build anything meaningful, they needed a common language and a rhythm they could count on. Agile offered the mindset shift, but Scrum offered the playbook. It wasn't about rules for rules' sake. It was about creating clarity in chaos. As Maya gathered the team around a blank whiteboard and a simple framework, she taught them Scrum and helped them see how structure could create freedom, not friction.

Team Vertex left the "Why Agile?" session with more questions than answers, but at least now, those questions felt hopeful. The solution was unclear, but the necessity of structure for the chaos was clear as day.

Maya gathered the team in a small, echoey conference room that still smelled like dry-erase markers and old ambition. "If Agile is the mindset," Maya began, "then Scrum is the rhythm. It gives us just enough structure to stay aligned without slowing us down."

The promise was clear:

"Let's keep this simple and useful."

SCRUM IN A NUTSHELL: THE 3-5-3 METHOD

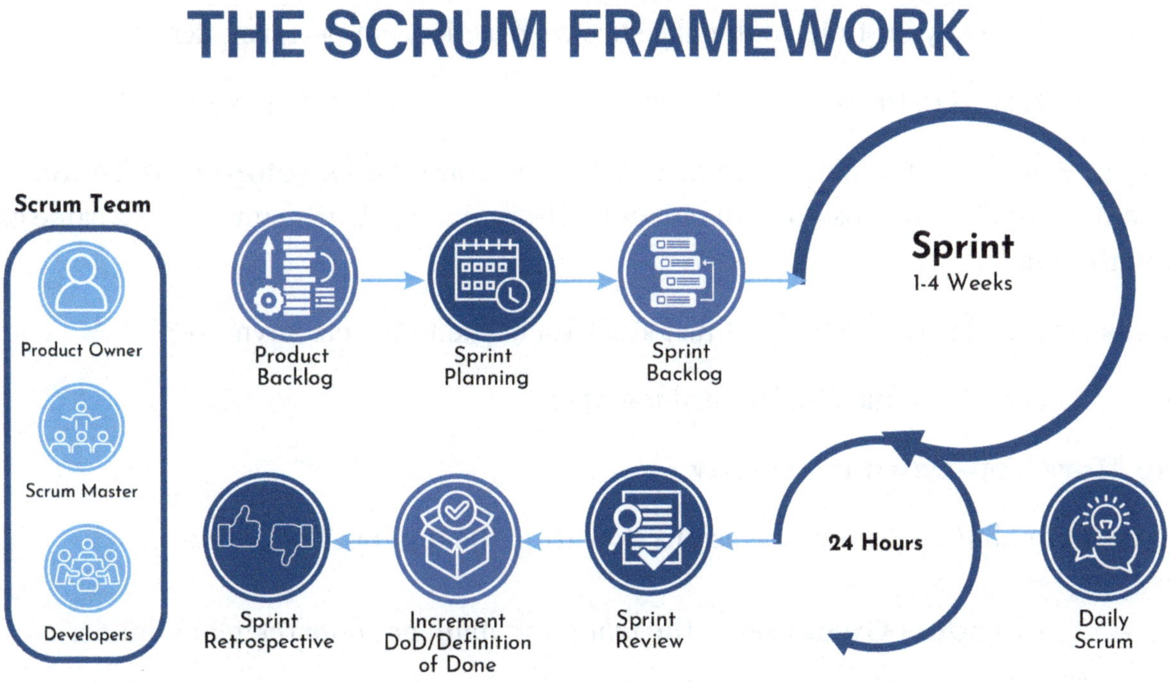

This visual represents Scrum as a rhythm, not a checklist. Everything revolves around the Sprint, with clear roles, events, and artifacts working together to create focus, feedback, and forward momentum. Rather than treating these elements as standalone concepts, Scrum relies on their interaction to help teams learn quickly and deliver value consistently. The next sections break each part down so you can understand not just what they are, but why they matter.

THE THREE SCRUM ROLES

So…Who Does What Around Here? The 3 Roles

> As roles began to solidify, Maya paused the meeting to make things unmistakably clear on who owns what, and why it matters…

"Maria, you're officially the **Product Owner** now. You already own the roadmap, but now you'll own the *what* and the *why*. You get to say, 'Here's what matters most.'"

Maria smirked. "Wait, does this mean I get to say no to random feature requests mid-Sprint?"

Maya smiled. "Exactly. You're the voice of the customer and the Product Backlog's guardian." She turned to the rest of the team. "I'm the **Scrum Master**, not your boss or project manager. I don't assign tasks, write schedules, or track performance. I remove blockers, protect your time, and make sure we practice, not perform, Scrum."

Then she turned to the board and drew a circle around the other team members.

Maya: "And the developers, testers, and designers are the **Development Team**. You own the *how*. You're cross-functional and self-organizing. One team with the same goal and rhythm."

Malik raised a brow. "So, I'm not just stuck on backend tickets anymore?"

Maya: "Nope. You build and succeed together."

No More Top-Down Hierarchy

In traditional project teams, decisions flow from the top. But in Scrum:

- The **Product Owner** owns the *what* and *why,* not *how* the team builds.
- The **Scrum Master** guides the *framework,* not people.
- The **Development Team** is *self-organizing and* decides the *how* together.

No one's a taskmaster. Everyone owns the outcome.

THE FIVE SCRUM EVENTS

Okay, But What *Actually* Happens in Scrum? The 5 Scrum Events

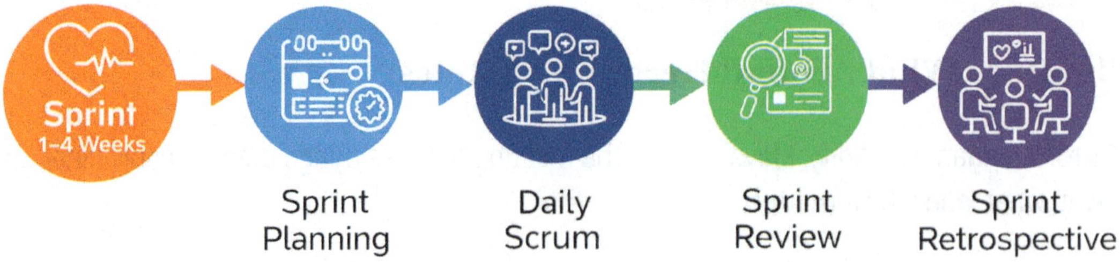

With the team still finding their footing, Maya stepped in to break down the five Scrum events, highlighting what they are and why they matter...

"Let's talk rhythm," Maya continued.

1. **Sprint:** "This is our heartbeat. We have two weeks. By the end, we deliver something real."
2. **Sprint Planning:** "At the start, we decide what we'll deliver within the two weeks and how we'll get there."
3. **Daily Scrum:** "Fifteen-minute daily sync. No status updates. The focus stays on progress and pivots."
4. **Sprint Review:** "We demo what we built. Stakeholders give feedback, and we adjust."
5. **Sprint Retrospective:** "Our space to reflect, grow, and speak honestly."

Ceremony	**Timeboxed Duration**	**Frequency**	**Who Participates**	**What Happens**	
Sprint Planning	2–4 hours (for 2-week Sprint)	Once at the start of each Sprint	Scrum Team (PO, SM, Devs)	Team discusses the Sprint goal, selects items from the backlog, and commits to work.	
Daily Scrum	15 minutes	Daily	Scrum Team	Each team member shares what they did, what they will do, and any blockers.	
Backlog Refinement (not an official Scrum Ceremony)	1–2 hours	1–2 times per Sprint	Scrum Team + Stakeholders	Team reviews upcoming backlog items, refines stories, and adds details or estimates.	
Sprint Review	1–2 hours	Once, at the end of each Sprint	Scrum Team + Stakeholders	Team demos completed work to stakeholders and gathers feedback.	
Sprint Retrospective	1–1.5 hours	Once, at the end of each Sprint	Scrum Team	Team reflects on the Sprint, discusses what went well, what didn't, and how to improve.	

Maya sketched three boxes on the board:

- **Product Backlog:** Owned by Maria. It's everything we *can* build.
- **Sprint Backlog:** "What will we build this Sprint. Pulled from the Product Backlog."
- **Increment:** "The working product we deliver. It doesn't have to be flashy, but it must work."

She circled the entire 3-5-3 framework.

"These are guardrails that keep us aligned without locking us in."

The 10-Step Scrum Cycle

Once the team had a handle on the roles, events, and artifacts, Maya leaned in."Knowing the pieces is one thing," she said. "But Scrum is more a rhythm than a checklist."

She drew a circle around the 3-5-3 diagram and added arrows looping from one event to the next. "No more linear patterns. This is a cycle we repeat. Every Sprint is targeted for growth."

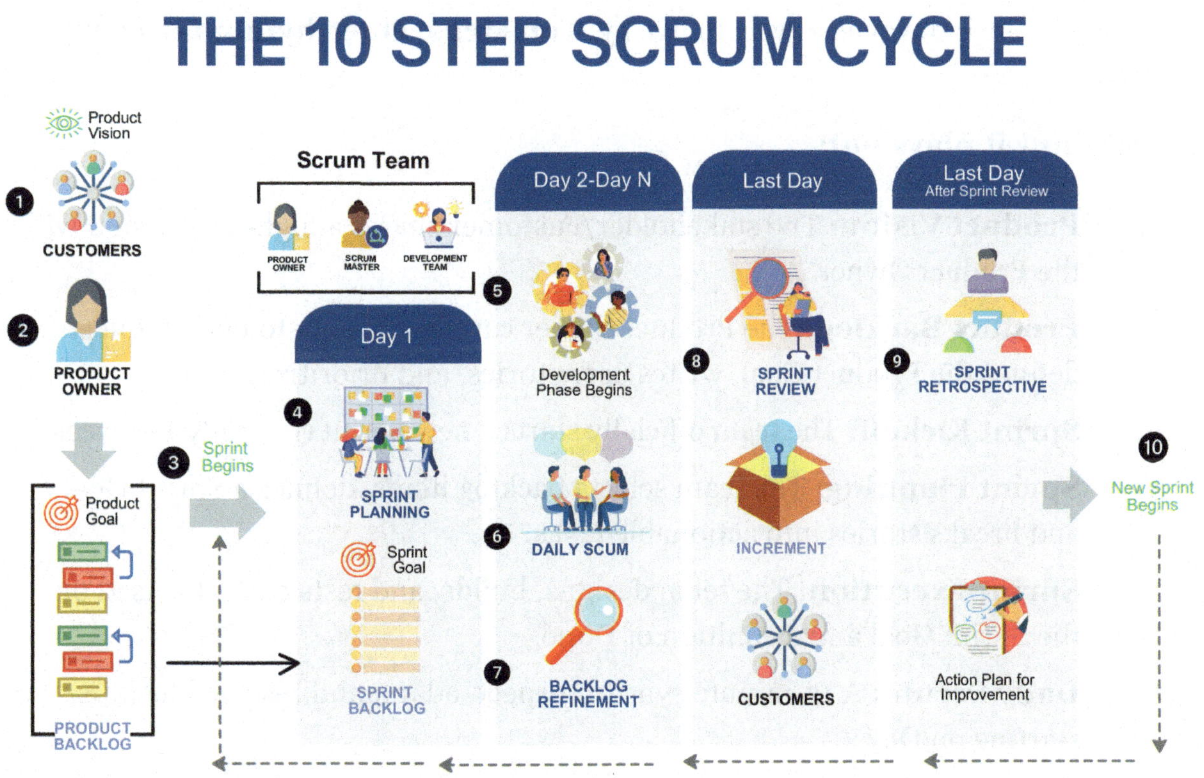

Then she turned to the whiteboard and wrote: "What Scrum actually looks like in motion."

Scrum Sprint Cycle: Where Structure Meets Flow

Team Vertex was in desperate need of a dependable rhythm they could move to even when priorities shifted or blockers crept in.

That's where Scrum delivered.

Maya drew a loop on the whiteboard.

"Scrum helps us find a repeatable flow so you can build smarter, adapt faster, and stay sane while doing it."

Than she broke it down as a cycle to live by. **Ten steps. One rhythm for every Sprint.**

Here's how it plays out:

1. **Product Vision:** The stakeholder/customer shares a high-level vision with the Product Owner.
2. **Product Backlog:** The Product Owner captures the customer's requests, defines the Product Goal, writes user stories, and prioritizes items.
3. **Sprint Kickoff:** The team officially starts a new Sprint (typically 1–4 weeks).
4. **Sprint Planning:** The team selects backlog items, defines a Sprint Goal, and breaks stories into actionable tasks.
5. **Sprint Execution:** The team designs, builds, and tests daily, focused on the Sprint Goal and Definition of Done.
6. **Daily Scrum:** A 15-minute sync to inspect, adapt, and resolve blockers starting on Day 2.
7. **Backlog Refinement:** Mid-Sprint, the team revisits and preps the backlog for upcoming Sprints.
8. **Sprint Review:** The team demos completed work, gets customer/ stakeholder feedback, and updates the backlog.
9. **Sprint Retrospective:** The team reflects on how they worked together and how to improve next time.

10. **Next Sprint Begins:** With lessons learned and priorities set, the next cycle begins.

Team Vertex First Assignment

Maya handed out sticky notes labeled with each Scrum role. There was hesitance, some excitement, and a few sideways glances. However, by the end of the meeting, Maria had fully stepped into the Product Owner role as a strategist and a voice of focus. The developers, including frontend, backend, QA, and design, agreed to operate as one cross-functional team. Also important: Maya was building momentum.

What Scrum Isn't

Before the team left, Maya dropped one final truth: "Scrum won't fix dysfunctions, but it'll make them visible. And visibility is where the real work begins." She clicked the cap back on her marker. "Scrum isn't micromanagement or just stand-ups and stickers. And it sure as hell isn't a magic fix."

The team laughed. But they also nodded. Deep down, they knew **what they had been doing before wasn't working.** Now, they had a rhythm to try.

By the end of the session, the roles were claimed, the events were named, and the artifacts were no longer mysteries. But more than knowledge, Team Vertex left with a sense of shared rhythm. Scrum wasn't about perfection. It was about progress with intention. They knew blockers would come, priorities would shift, but with 3-5-3 as their foundation, they had a way to navigate the uncertainty together.

What's Next? Applying Scrum to the Full Delivery Lifecycle

Now that Team Vertex had a grasp on Scrum's framework, the next challenge was understanding how to apply it across the entire product lifecycle. Scrum is a set of rituals and a delivery engine for real software. In the next chapter, we'll explore how the traditional SDLC maps to Agile workflows, and how Scrum helps teams build, test, and ship value one Sprint at a time.

Exercise: Match the Role, Own the Outcome

Team Vertex just learned about the 3 Scrum roles, 5 events, and 3 artifacts. Apply what you've learned by matching the responsibilities below to the correct Scrum component.

PART 1: MATCH THE ROLE

Match each responsibility to **Product Owner**, **Scrum Master**, or **Development Team**:

Responsibility	Scrum Role
A. Removes blockers and facilitates team meetings	
B. Writes and prioritizes user stories	
C. Builds, tests, and designs the product	
D. Guards the Sprint from external disruptions	
E. Owns the Product Backlog	

PART 2: IDENTIFY THE EVENT

Which Scrum event fits each purpose?

Purpose	Scrum Event
A. Inspect and adapt the working product with stakeholders	
B. Set the goal and scope for the next Sprint	
C. Hold a quick daily sync on progress and blockers	
D. Reflect and improve on how the team worked	
E. Time-boxed period to build an Increment	

PART 3: ARTIFACTS IN ACTION

For each description, write the corresponding **Scrum Artifact**:

Description	Artifact
A. The source of truth for everything we could build	
B. What we plan to deliver in this Sprint	
C. The usable product outcome of a Sprint	

6. NAVIGATING SDLC IN SCRUM

In traditional software development, the Software Development Life Cycle (SDLC) unfolds as a linear sequence from planning to deployment, often requiring full completion of one phase before the next begins.

But in Scrum, the SDLC is not discarded; it's reimagined. Rather than following a rigid waterfall model, Scrum teams cycle through SDLC phases continuously within each Sprint, allowing for faster feedback, better alignment, and more frequent delivery of working software. This chapter breaks down how SDLC activities map to the Scrum framework, showing how initiation, planning, design, development, testing, and release all take place in iterative, collaborative loops.

Scrum is more flexible than the traditional approach to project planning and execution. In a Scrum environment, the Software Development Life Cycle (SDLC) looks like this:

SDLC PHASES MAPPED TO SCRUM ACTIVITIES

SDLC Phase	How It Happens in Scrum	Related Scrum Event / Practice
1. Initiation & Planning	Define product vision, high-level goals, and begin backlog creation	Sprint Zero, Product Vision Workshop, Backlog Setup
2. Requirements Analysis (Backlog Refinement)	Break down epics into stories, clarify AC, estimate, and prioritize	Backlog Refinement Sessions
3. Design	Just-in-time design for Sprint stories, including wireframes, UX flows, and technical spikes	Sprint Planning, In-Sprint Design Collaboration
4. Development	Code implementation of user stories during the Sprint	Sprint Execution
5. Testing	Write and execute tests during the Sprint; verify stories meet AC and Definition of Done	Sprint Execution, QA Testing
6. Deployment & Release	Deploy potentially shippable increments at Sprint end via CI/CD pipelines	Sprint Review, Release Planning
7. Maintenance & Retrospective	Log bugs, gather feedback, and reflect on process improvements for future Sprints	Sprint Retrospective, Backlog Updates

By integrating SDLC phases into Scrum's recurring Fast Feedback loop, teams deliver working software every Sprint, keep quality high, and adapt quickly to new information rather than waiting for a lengthy, linear development process to complete.

CLOSING REFLECTION

By integrating SDLC phases into Scrum's iterative engine, teams preserve the discipline of structured software development while gaining the flexibility to respond to change. Each Sprint becomes a complete micro-cycle of delivery, from planning to release, with learning and adaptation built in. This approach not only shortens time-to-value but also raises quality, transparency, and alignment across the team. When teams stop treating SDLC as a checklist and start treating it as a rhythm, they deliver more than software. They deliver outcomes *every* Sprint.

Exercise: SDLC in Motion

INSTRUCTIONS:

Match each Scrum activity below to the **correct SDLC phase** (based on the 7-phase model used in this chapter). Then, explain briefly *why* it fits that phase.

SDLC Phases to Use:

1. Initiation & Planning
2. Requirements Analysis
3. Design
4. Development
5. Testing
6. Deployment & Release
7. Maintenance & Retrospective

SCRUM ACTIVITY EXAMPLES

Scrum Activity	SDLC Phase	Why it fits
A. Sprint Planning meeting		
B. Writing user stories and clarifying AC		
C. Developers writing and testing code during the Sprint		
D. Design team creating mockups for the current Sprint stories		
E. Holding a Sprint Retrospective to reflect and improve		
F. Demoing finished work in Sprint Review and preparing for release		
G. Identifying bugs post-release and adding them to the backlog		

REFLECTION PROMPT

- How does cycling through SDLC phases inside each Sprint help Agile teams adapt faster than in traditional models?
- Which SDLC phase do you think is most vulnerable to being rushed or skipped in Scrum...and what could help prevent that?

7.

SPRINT 0: BUILDING THE FOUNDATION

SPRINT ZERO: KEY DEFINITIONS AND PRACTICES

DEFINITION OF READY (DOR)

The **Definition of Ready (DoR)** is an agreed-upon checklist of criteria that a user story or product backlog item must meet before the team accepts it into a Sprint. In Agile practice, this ensures each requirement is fully understood, refined, and actionable. For example, having a clear description, defined business value, acceptance criteria, and resolved dependencies *before* development begins.

A strong DoR prevents incomplete or ambiguous backlog items from entering a Sprint, thereby improving clarity and reducing mid-Sprint scope changes. It also supports effective Sprint Planning: only items that satisfy the DoR are pulled into the Sprint, meaning the team can confidently commit to deliverables that are ready for implementation. This shared understanding of "ready" between the Product Owner and Development Team helps align expectations and sets the stage for a smoother execution in each Sprint.

DEFINITION OF DONE (DOD)

The **Definition of Done (DoD)** is a formal set of completion criteria that each product increment or deliverable must fulfill to be considered truly "done." It is essentially a quality checklist agreed upon by the team, encompassing all activities needed to deliver a complete and usable outcome. In design contexts, the DoD ensures that all design artifacts (such as wireframes, mockups, or specifications) are finished to the agreed standard, for instance, reviewed, approved, and meeting usability guidelines, before calling the design work complete. This guarantees completeness and usability of the design outputs, not just their functionality, so that the development team has everything required to proceed with coding and implementation. In other words, when a design item is "done" under this definition, developers can build with confidence and testers can verify against a fully defined design, ensuring no critical elements are missing for a potentially shippable product increment.

ACCEPTANCE CRITERIA IN GHERKIN FORMAT

Acceptance Criteria in Gherkin format are written as scenario-based statements using the **Given-When-Then** syntax, a style originating from Behavior-Driven Development (BDD). In this format, each user story's expected behavior is defined through concrete examples:

Given some initial context or precondition,
When an action is performed,
Then an expected outcome should occur.

Writing acceptance criteria in the Given/When/Then structure makes requirements highly testable and unambiguous. Each scenario functions as a clear, testable condition that tells the team exactly what to implement and the tester exactly what to verify. In fact, Gherkin-formatted criteria help testers determine when to begin and end testing for a story, and they often reduce the time needed to create test cases by describing system behavior upfront. This format is used for its precision and clarity: it creates a shared understanding across Agile teams by communicating requirements in a consistent, human-readable way.

Developers, testers, and Product Owners all interpret the criteria the same way, which improves alignment and ensures that everyone knows how to measure the success of a story. Overall, using Gherkin-style acceptance criteria promotes transparency and helps the team verify that a user story meets the intended "Done" when the conditions are fulfilled.

WIREFRAMES & PROTOTYPES (DESIGN)

Wireframes and prototypes are essential design artifacts used to visualize user stories and product features before development commences**. Wireframes** are typically low-fidelity sketches or outlines of user interface layouts. They serve as a communication tool and an early design blueprint, focusing the direction of the design in the project's initial stages (often during Sprint Zero).

Creating wireframes allows the Product Owner, designers, and developers to collaborate on the envisioned solution, while the development team refines requirements and user stories, designers draft wireframes to depict the proposed screens and interactions. This process invites feedback from both the team and stakeholders, helping to confirm the project's "focus" and ensuring everyone shares the same mental model of the feature before any code is written.

Once the basic layout and approach are agreed upon via wireframes, designers progress to building **prototypes.** Prototypes are more refined, interactive models of the product (ranging from click-through screen mockups to higher-fidelity simulations) that demonstrate how users will interact with the system. The key goal of prototyping in an Agile context is to continue gathering feedback and validating the user experience through iterative improvements. By testing prototypes, the team can identify usability issues or misunderstandings early and adjust the design accordingly, all before development locks in the implementation.

In summary, wireframes and prototypes in Sprints foster a shared understanding between the Product Owner, UX/UI designers, and developers. They act as visual specifications of the user story, aligning the team on how the feature should look and behave. This significantly reduces the risk of misinterpretation during development by ensuring the design intent is clear and agreed upon upfront.

DEVELOPMENT: ENVIRONMENT SETUP & TECHNICAL ENABLEMENT

In Sprint Zero, development isn't about building features – it's about enabling the team to build without interruption once Sprint 1 begins. Teams establish the foundational technical environment so developers can code, integrate, and test safely and consistently.

Key focus areas include: **Repository creation & branching strategy** (e.g., main, develop, feature branches), **Local dev environment setup** (frameworks, libraries, mobile emulators, etc.), **Continuous Integration (CI) kickoff** – automated builds begin now, and **Core architecture decisions** and initial proof-of-concept spikes for risky components

Rather than writing production code, Sprint Zero aggressively removes development blockers. The goal: the team never stops in Sprint 1 because Git isn't set up or software licenses are missing. The technical foundation created here accelerates delivery in every future sprint.

TESTING: QUALITY & AUTOMATION BASELINE

Testing in Sprint Zero also focuses on readiness rather than validating features. The QA mindset starts on Day 0 – not once code appears. Core testing deliverables include: Test strategy & tooling selection (functional, accessibility, performance, security), Shared quality standards aligned with the Definition of Done, Automation framework setup – even a "hello world" test running in CI execution, and Traceability structure between acceptance criteria → test cases → pipelines

This early alignment ensures that every user story in Sprint 1–N arrives testable by design, and that quality becomes a habit.

DEPLOYMENT/RELEASE: FIRST PATH TO PRODUCTION

There is no production release in Sprint Zero — but the team builds the delivery pipeline that will get them there later. The objective is to establish a repeatable path so deployments are routine and boring. Sprint Zero deployment goals:

- CI/CD pipeline configured (build → test → package → deploy)
- Knowledge of environments and their purpose:

    - DEV = experiment freely

    - QA = integrated / validated

    - UAT = stakeholder/business acceptance

    - PROD = real users, real stakes

- Access, credentials, and permissions resolved upfront
- Rollback strategy drafted (we plan for failure while planning success)

This reduces late-stage surprises, such as "no one can push to UAT" the night before release.

MAINTENANCE & RETROSPECTIVE: CONTINUOUS IMPROVEMENT INFRASTRUCTURE

Even before a single feature is built, Sprint Zero lays the groundwork for continuous learning and adaptability — the foundation of agility. Key maintenance and inspection-adaptation setup:

- **Issue logging structure** for bugs and technical debt
- **Monitoring + telemetry approach** (e.g., performance metrics, crash analytics)
- **Retrospective cadence and tools** defined (Miro, Confluence templates, etc.)
- **Working Agreements established** — how we communicate, collaborate, and improve

The retrospective for Sprint Zero isn't about delivering product — it's about how ready we are to deliver product and what still blocks us.

Sample retrospective questions: What slowed us down during Sprint Zero that could sabotage feature delivery? How confident are we in delivering a Done increment in Sprint 1? Which risks need mitigation in the next 2 weeks?

Sprint Zero exists only when the team is truly ready to inspect and adapt through working software — not words.

BRING THE ARC TOGETHER

SDLC Phase	Sprint Zero Output
Planning	Vision, roadmap, prioritized backlog
Requirements	Acceptance Criteria + Gherkin, INVEST-ready stories
Design	Wireframes & prototypes, UX alignment
Development	Environments, repos, CI setup
Testing	Test strategy + automation baseline
Deployment/Release	CI/CD pipeline + path to environments
Maintenance/Retrospective	Working agreements, issue logging, improvement loop

The finish line of Sprint Zero is the ability to ship starting Sprint 1.

Sprint Zero isn't about producing shiny artifacts or pretending progress has been made — it's about eliminating every excuse that would prevent real delivery once Sprint 1 begins. By aligning on readiness, quality, technical capability, and how the team will continuously improve, we remove friction before it ever reaches the work. Now that the foundation is in place, Team Vertex is ready to build working software that can actually be shipped. The real adventure starts in Sprint 1, where ideas stop living on whiteboards and begin transforming into features in users' hands.

SPRINT 0

PHASE 1:
PLANNING
PHASE 2:
REQUIREMENTS
PHASE 3:
DESIGN
PHASE 4:
DEVELOPMENT
PHASE 5:
TESTING
PHASE 6: DEPLOYMENT
PHASE 7:
MAINTENANCE

8. SPRINT 0: PHASE 1

PLANNING PHASE: PRODUCT VISION AND RELEASE STRATEGY

INTRODUCTION

Before a single feature could be built, Team Vertex had to ensure the ground they were standing on was stable. Sprint 0 wasn't just orientation, it was a rehearsal, a reset, and a readiness test all in one. Although no working product increment would be delivered to end users, the team moved intentionally through all seven phases of the Software Development Life Cycle (SDLC). From defining the product vision to deploying a navigable app shell, Sprint 0 served as a complete mini-cycle. By the end of it, they weren't just a group of professionals assigned to a backlog; they were an **aligned and enabled team**. The tools were connected, the backlog was groomed, the designs were shared, and the pipeline was alive. Sprint 0 proved that Team Vertex wasn't guessing, **they were ready**.

OBJECTIVE

Sprint 0 teaches the crucial steps of project initiation. The goal was to establish a shared understanding and set up everything needed for development. In this Planning/Initiation phase, Team Vertex focused on aligning stakeholders around a clear product vision, creating key planning artifacts (like scope definitions, stakeholder maps, and risk plans), and preparing the team and environment for Sprint 1. This foundation ensured the Scrum Team would begin Sprint 1 with clarity, infrastructure readiness, and a well-groomed initial backlog.

Feature in Focus: Product Vision and Release Planning

Sprint 0 centered on defining **what** the team would build and **how** they would deliver it. By the close of this planning phase, Team Vertex had a concrete plan in place. In fact, Sprint 0 produced a suite of outcomes that set the stage for success:

- **Product Vision:** A shared understanding of *why* the product exists and *who* it serves, encapsulated in a clear, outcome-oriented vision statement.
- **Agile Project Scope:** An outline of what will (and won't) be delivered in the MVP, providing boundaries for the initial release.
- **MVP and MMP Definition:** Clarity on the Minimum Viable Product vs. next-phase enhancements, so everyone knows which features are now or later.
- **Stakeholder Map:** Key players identified and their roles in the project collaboration defined, ensuring the right people stay involved at the right level.
- **Risk Radar:** Early risks surfaced along with mitigation strategies, so no one is caught off guard when development begins.
- **Release Planning:** A high-level planning aligned with team velocity, phasing the rollout into manageable launches rather than one big bang.

Together, these planning outputs gave the team and stakeholders a clear vision and strategy. Sprint 0 ended with everyone on the same page about *what* they were building and *how* they'd approach delivering it.

TEAM VERTEX IN ACTION

Enter PetTracker, Team Vertex's new project and the focus of Sprint 0. PetTracker is a mobile application designed to help pet parents monitor their pets' health and routines in real time. Planned features included logging meals, walks, and medications, setting care reminders, tracking vet appointments, and monitoring pet behaviors. This project served as the throughline example for each SDLC phase, starting with the Planning/Initiation groundwork laid in Sprint 0.

Day 1: Crafting a Shared Vision

At the very start of Sprint 0, Maria (Product Owner) gathered the team for a project kickoff session. Before anyone wrote a single user story or opened a whiteboard for Sprint Planning, Maria did something crucial: she stepped back and asked *why* they were building anything at all. She led the team in defining a clear, outcome-oriented product vision. Maria proposed an initial vision statement for PetTracker:

> "Empower pet parents to track meals, walks, and medications so they never miss a beat in their pet's care."

This simple, user-focused declaration became their North Star, guiding feature prioritization and user-centric decisions throughout Sprint 0. To ensure everyone was truly aligned, Maria wrote the vision on a shared board and posed a question to the team: **"What would success look like if we nailed this in 3 months?"**

That brief vision ceremony grounded every decision that followed from persona mapping to MVP scoping by uniting the team around a common purpose.

SIDE NOTE: "IF YOUR TEAM CAN'T REMEMBER THE PRODUCT VISION WITHOUT READING IT, IT'S TOO LONG OR TOO VAGUE. KEEP IT SHORT, EMOTIONAL, AND USER-FOCUSED."

Bad vs. Good Product Vision Examples

To reinforce the importance of a strong vision, the team reviewed examples of weak vs. improved product visions:

Bad Vision	Why It Fails	Improved Vision
"Be the #1 app for pet care."	Too vague; focuses on market rank, not on a user problem.	"Help pet parents track daily routines so their pets stay healthy and happy."
"Disrupt the scheduling space using advanced technology."	Buzzwordy; no clear user or problem stated.	"Make it easy for service pros to book and manage appointments from their phone."
"Innovate in mobile wellness."	Empty language; no audience, problem, or purpose.	"Give wellness coaches a simple way to track client progress and habits on the go."
"Provide a seamless platform that leverages synergy."	Corporate fluff; unmemorable and uninspiring.	"Help freelancers get paid faster with fewer steps and less stress."

A well-crafted vision inspires action. For Team Vertex, aligning on the "why" behind PetTracker gave meaning to the work ahead and clarity to the product's direction. The cross-functional team rallied around this vision, ensuring every feature they planned to build would serve a real user need.

Day 2: Co-Creating the Agile Project Scope

Scene: Team Vertex Assembles (Day 2 of Sprint 0). The newly formed team gathers in a breakout room plastered with sticky notes. There's a wave of uncertainty; everyone has been pulled from different departments, and they're still forming as a unit. Maya (Scrum Master) opens the session:

Maya: "We align before running. Let's co-create our Agile Project Scope."

In Agile, *scope* isn't a 50-page specification document, it's a **shared understanding**. The Agile Project Scope provides just enough structure to align the team on *what* they're building, *why* it matters, *who's* involved, and *how* they'll work together, without overplanning or locking down details too early. Team Vertex collaborated to outline key elements of their project scope, covering everything from the product vision and objectives to roles, deliverables, and working agreements.

They identified the **product's objective statement** (deliver an MVP that ensures no pet care activity is missed), listed out MVP **deliverables** versus deferred features, clarified **stakeholders** and core team roles, set up a rough **schedule** (1-week Sprint 0, followed by four 2-week Sprints for MVP delivery), and even agreed on basic **working agreements** (daily standups at 10 AM, Slack for async communication, etc.).

They defined what "*Done*" means for any story (code merged, peer-reviewed, tested, and accepted by the PO) and what conditions must be met for a story to be ready for development. By the end of Day 2, the team had built a one-page Agile Project Scope document that captured the boundaries and expectations for the PetTracker project.

Closing Reflection: With the Agile Project Scope defined, Team Vertex had more than a plan, **they had alignment**. Everyone understood the purpose, boundaries, roles, and expectations from the start. This foundation didn't lock the team into rigid commitments; it gave them the clarity to inspect and adapt with confidence. In Agile, scope is about building shared understanding before momentum begins.

PROJECT SCOPE

Project Name: PetTracker Mobile App, v1.0

Project Sponsor: Victor, VP of Product

Project Date: 07/11/2026

Product Owner: Maria

Scrum Master: Maya

Project Vision

To help busy pet owners simplify and stay on top of their pets' daily care through real-time tracking, reminders, and digital health logs, so pets stay healthy and owners feel in control.

Objective

To deliver a user-friendly mobile app that empowers busy pet owners to manage daily care routines—such as feeding, walking, and medication tracking—through real-time logs, smart reminders, and digital health records, improving pet well-being and owner confidence.

Stakeholders

- VP of Product - Project Sponsor
- Customer Support Lead - Feedback loop from users
- Lead Vet Consultant - Domain expert
- Compliance Offer - Risk/goverance
- CTO - Tech alignment
- Pet Parent - End user
- Pet Trainers - End User

High-Level Scope

In-Scope (MVP):

- Real-time pet activity logging
- Feeding and medication reminders
- Profile setup for multiple pets

Out-of-Scope (Phase 1):

- Smart home integrations
- Lost pet community alerts
- Social sharing or community forums

Deliverables

A functional, cross-platform mobile application (iOS & Android) that enables pet owners to track feeding, walks, medication, and health history through real-time logging, reminders, and digital records—with a user-friendly interface and MVP-ready feature set.

MVP Deliverables Include:

1. User Registration & Login
2. Create & Edit Pet Profile
3. Log a Meal (with portion type: wet/dry/raw)
4. Set Feeding Reminders
5. Track Walks (Start, Stop, Notes, GPS permission)
6. Basic Medication Log (name, dosage, time)
7. Basic Pet History View (logs by type/date)
8. Error Handling & Validation (forms/screens)
9. UAT-Approved MVP Build (Downloadable)
10. Agile Documentation: Backlog, Wireframes, Acceptance Criteria

MMP Deliverables Include: (Enhanced Functionality)

1. Recurring Reminders & Push Notifications
2. Multiple Pet Support (add/edit/remove pets)
3. Pet Health History Export (PDF/email)
4. Medication Reminders & Alerts
5. Walk Tracking History & Duration Analytics
6. UX Enhancements (profile photos, color themes)
7. Improved Error Feedback (toasts, modals)
8. Offline Mode & Data Sync
9. Beta Feedback Collection Tool

Success Criteria

- MVP available in app store by end of Q2
- 90%+ user retention after 30 days
- <5% bug rate in production
- 15% upgrade to premium within 6 months

PROJECT SCOPE

Constraints & Dependencies

- Integration with third-party GPS APIs
- Regulatory compliance for pet data (if applicable)
- Team availability: 2 developers full-time & 1 part-time

Roles & Strengths

- Alex (Dev)
- Jenna (UI/UX Designer)
- Malik (QA)
- Liam (API Dev)

Working Agreements

- Sprint length: 2 weeks
- Sprint Planning: (Wk1) Wednesdays 9:30 AM
- Backlog Refinement & Estimation: (Wk2) Wednesday 9:30 AM, Sprint Review: Based on stakeholder availability
- Daily Scrum: 10 AM EST
- Figma for design reviews
- Clickup for product & sprint backlog
- Slack for async check-ins
- GitHub for version control & pull request
- Notion for documentation

Team Purpose

To collaboratively deliver high-quality, user-centered solutions that align with the product vision, continuously improve through feedback, and foster a culture of transparency, ownership, and cross-functional collaboration.

Definition of Ready

- Story has clear AC that are testable
- Linked to Product Goal
- Prototype, wireframe, and mockup are linked
- Required backend/API contract are known
- Dependencies identified
- Sized appropriately by team
- Story are marked "Ready" in the board

Definition of Done

- Code merged, peer reviewed, and check into the correct branch
- All tests completed and passed
- Story Demoed in Review
- Accepted by PO
- UX & design align with features
- All tasks/subtasks are closed
- Documentation is updated
- Story status is moved to "Done" on the board

Day 3: Stakeholder Mapping & Engagement

Scene: Team Vertex Faces a Wall of Opinions (Day 3 of Sprint 0)

Maria (PO) walks into the planning meeting looking flustered. "I've had five different VPs ping me about PetTracker," she sighs. "One wants a dashboard, one wants community features, and one just sent a spreadsheet of 'must-have' functionality."

Maya nods in agreement: "We need a stakeholder strategy. Otherwise, this MVP's going to collapse under shifting demands."

Before the team gets pulled in every direction, they pause to do **Stakeholder Mapping**. This practice helps Agile teams identify, categorize, and figure out how to engage the people who influence (or are affected by) the product.

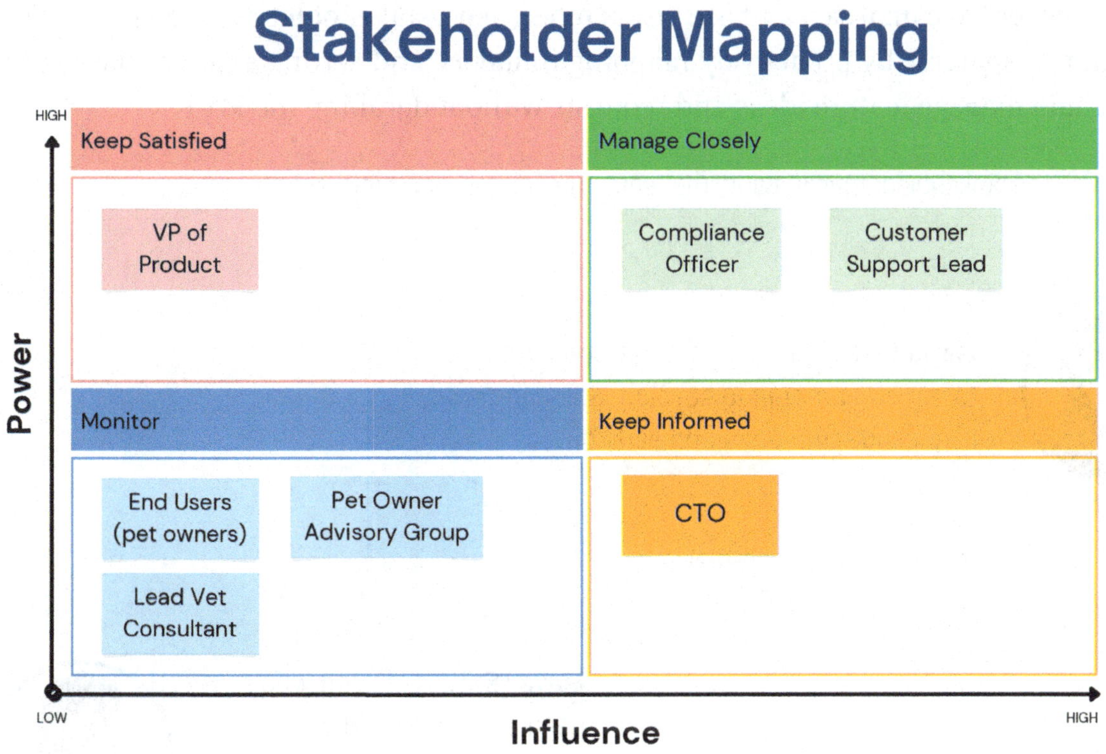

In an Agile environment, where change is constant, stakeholder *clarity* is power. Not all stakeholders play the same role or carry the same weight, so Team Vertex first brainstorms the types of stakeholders involved in PetTracker: from **Champions** (executive sponsors backing the product) and **End Users** (pet owners, vets who will use the app), to **Cross-Functional Partners** (Marketing, Compliance, Support), **Technical Enablers** (Security, DevOps), and even **Peripheral Influencers** (folks in Finance or Legal who have tangential interest).

With a clear view of "who's who," Maria facilitated a mapping session to drill down into specific names and needs. The team listed key stakeholders, each one's role or interest, their level of influence on the project, and an appropriate engagement strategy for each. For example, the **VP of Product** (project sponsor) had high influence, so they planned weekly check-ins and roadmap reviews to keep him satisfied. A **Customer Support Lead** would provide user feedback with medium influence, so inviting them to Sprint Reviews made sense. A **Lead Vet Consultant** (domain expert) would be engaged on an ad-hoc basis for input on pet health features. The **Compliance Officer** (high influence on governance) would be involved early for regulatory review. Not everyone would get a seat at every demo. The team defined a cadence for who gets what information, how often, and in what format.

They also agreed on some **engagement best practices**: co-create priorities but let the Product Owner have final say, and use visual tools (like a stakeholder radar chart) to keep this plan transparent. By proactively mapping influence and setting expectations, Team Vertex could manage feature ideas and requests without derailing the MVP.

With the stakeholder map in hand, the team immediately put it into action:

Maria (PO): "I'll invite the Vet Consultant to our story mapping session. She can help us identify any gaps early on."

Maya (SM): "Let's log who's been consulted and who hasn't. I'll set up a shared stakeholder board in Miro for us to track engagement."

Malik (QA): "If Compliance only comes in at the end, we'll be testing under fire. Let's pull them in early to review requirements."

Stakeholder Engagement Plan

Stakeholder	Role/Interest	Level of Influence	Engagement Strategy
VP of Product	Project Sponser	High	Weekly check-in, roadmap reviews
Customer Support Lead	Feedback loop from users	Medium	Sprint Review attendance
Lead Vet Consultant	Domain expert	Medium	Invite to backlog refinement
Compliance Officer	Risk/governance	High	Early UAT involvement
CTO	Tech alignment	Medium	Monthly showcase/demo
Pet Advisory Group & Pet Owners	End user voice	Low	Quarterly survey + beta group

By involving the right people at the right time (and politely deferring the rest), the team protected the product scope from death by committee. They were learning to build with *stakeholders*, not for every stakeholder whim.

Day 4: Surfacing Risks Early

Team Vertex had learned the hard way that risk planning can't wait. "Last Sprint, a third-party API changed unexpectedly, breaking our GPS tracker just days before a demo. Everyone scrambled," Alex (Developer) reminded the group.

Alex: "We never saw it coming."

Maya: "Or we did and ignored it."

Determined not to repeat history, the team dedicated time in Sprint 0 to get ahead of potential pitfalls. Maya pulled up a blank whiteboard titled **"Risk Radar"**.

Maya (Scrum Master): "Let's surface what could go wrong before it becomes a blocker."

They set about brainstorming what *might* threaten their plan. In Agile Risk Planning, the goal is to continuously **identify, track, and mitigate** risks without getting bogged down in bureaucracy. Instead of a static 40-page risk register, Team Vertex kept a live risk board that they would update at each retrospective. They started by mapping common risk **categories** relevant to PetTracker, giving real examples for each to make them tangible:

- **Technical Risks:** e.g., third-party integrations could change unexpectedly (as they just experienced with the GPS API).
- **Delivery Risks:** e.g., timelines and team capacity. Jenna (Designer) mentioned she had PTO overlapping an upcoming Sprint, a potential hit to design bandwidth.
- **Requirements Risks:** e.g., scope creep or unclear acceptance criteria, like if a VP tries to add a new feature mid-Sprint.
- **Dependency Risks:** e.g., reliance on external teams or vendors. Their backend developer was shared with another team, a possible bottleneck.
- **Compliance Risks:** e.g., regulatory or privacy issues. Storing pet health data might invoke HIPAA-like rules.

For each risk identified, they immediately discussed a *mitigation plan*. For instance, to counter **API delays**, Alex suggested building mock endpoints so development wouldn't stall waiting on the real thing. To address cross-platform **GPS issues**, Malik proposed a quick spike (research experiment) in Sprint 0 and early device testing. To prevent **data duplication** errors, they'd implement validation patterns and have QA focus on edge cases. And to avoid **MVP scope creep**, Maria emphasized using **MoSCoW prioritization** (Must-haves, Should-haves, Could-haves, Won't-haves) and strict backlog hygiene. All these risks and responses were logged transparently in Jira, visible to the whole team, and they committed to reviewing the risk board before each Sprint Planning.

The discussion didn't stop at identification. How to respond was just as important. Malik wanted to know, "So what do we actually do when a risk materializes? Panic and pray?" Maya laughed and broke down the four classic risk responses:

Maya: "We have four moves: avoid it, mitigate it, transfer it, or accept it."

Malik: "Okay, break it down."

Maya: "Sure. If a risk looks too dangerous, we avoid it completely, for example, use an in-house geolocation module instead of a flaky third-party API. If we can't dodge it, we mitigate. Write early integration tests to catch API changes before they explode. If it's out of our hands, we transfer it. Say, we might rely on the platform provider's built-in security updates rather than building our own security layer from scratch, effectively transferring that risk to them. And sometimes, we just accept it, like App Store approval delays. We'll live with that risk and monitor it, but we won't lose sleep over it."

Malik: "Got it. Avoid, Mitigate, Transfer, Accept...Sounds like a risk management rap."

By the end of the risk session, Team Vertex had a short list of top risks with contingency plans for each. No one was crossing their fingers and hoping for the best, as they had already agreed on what to do *when* (not if) certain issues arose. This early risk planning boosted the team's confidence that they could handle surprises down the road.

Risk Response Plan

Risk Type	Description	Example	Response Strategy	Risk Response Plan
Technical	Code, architecture, integrations	API changes from GPS provider	Mitigate	Use in-house geolocation module instead of third-party API.
Delivery	Time, capacity, overcommitment	Jenna's PTO overlapping design sprint	Avoid	Adjust sprint schedule or reassign design tasks in sprint planning to ensure coverage.
Requirements	Shifting scope, unclear acceptance criteria	VP adding new feature mid-sprint	Mitigate	Enforce change control policy; defer changes to backlog refinement for next sprint.
Dependencies	External vendor, shared resources	Backend Dev shared with another team	Accept + Mitigate	Add buffer in sprint for shared resources; communicate early with dependent team.
Compliance	Legal, data, accessibility	Pet data storage must meet HIPPA privacy rules	Transfer + Mitigate	Consult legal/compliance experts; use third-party storage solutions with certifications.

Day 5: Agile Release Strategy

With a clear vision, scoped backlog, engaged stakeholders, and mitigated risks, the team turned to one final question: **How will we deliver value to users?** Rather than aiming for a big-bang release (which often leads to either champagne celebrations or chaos), Team Vertex embraced an Agile release strategy: paced, intentional, and feedback-driven. Each release phase would deliver *value*, not just code, and help build trust with users and sponsors.

Maya gathered the team for a release planning discussion. *"A release isn't just code going live,"* she reminded them. *"It's a promise to our users. Let's plan like it."* They all knew the MVP would be ready after a few Sprints, so the question "**When can we go live?**" was looming from leadership. The team wanted to avoid a risky all-or-nothing launch. Instead, they decided to **deliver in waves**:

Alex (Developer): "Let's break this into three phases: Internal Test, Friends & Family Beta, Public Launch."

Everyone agreed to this phased rollout. Maria (PO) immediately started tagging backlog stories by release phase:

- **Internal Test (Alpha):** Core logging features (meal logs, walk tracker, medication tracking) that the team would test in-house.
- **Friends & Family Beta:** Add-ons like reminder notifications and a summary dashboard. Once core features were stable, these would be released to a small external group for broader feedback.
- **Public Launch:** The remaining nice-to-haves and polish, e.g., a pet referral system, data export, and final UI refinements, which are to be delivered when the product goes fully public.

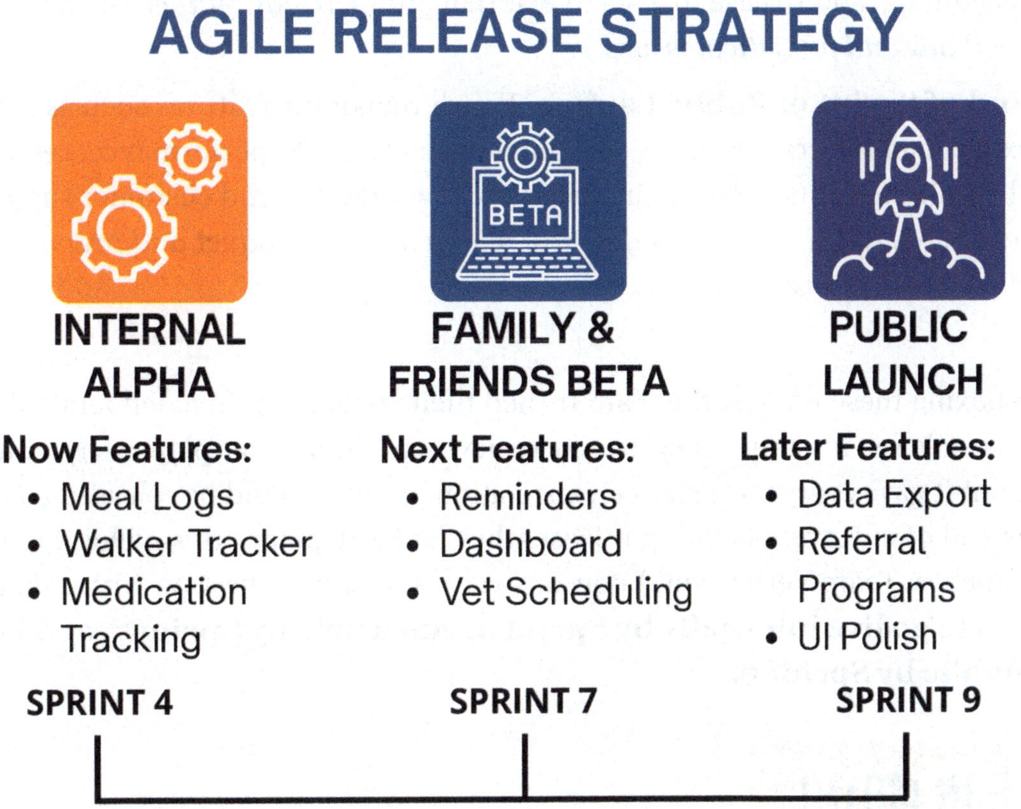

Maya checked their average velocity and raised a caution that became the team's mantra: *"This timeline only works if we buffer for testing, bugs, and feedback fixes."* In other words, each phase needed slack for quality assurance and iteration, not just development.

To make things concrete, Team Vertex wrote down definitions for each release phase. An **Internal Alpha** would be a private release to internal staff; a controlled environment to validate stability and basic workflows. The **Friends & Family Beta** would open the app to a select group of trusted external users (e.g., friends, partners, loyal customers) to focus on usability and catch any bugs in real-world use. The **Public Launch** would be the full release to everyone, only after performance was tuned, all major issues resolved, and support & marketing were in sync for go-live.

Finally, the trio of Maria, Alex, and Maya sketched out a high-level timeline tied to their Sprints:

- **End of Sprint 4: Internal Alpha**. By then, implement and stabilize all core logging features and have an internal demo for stakeholders. *(Goal: validate core functionality in a controlled setting.)*
- **End of Sprint 7: Friends & Family Beta**. Incorporate feedback from alpha, complete secondary features like reminders, dashboard, and vet scheduling, and release to a small external group. *(Goal: gather early user feedback and fix critical issues.)*
- **End of Sprint 9: Public Launch**. Finish remaining features such as data export and referral program, plus final polish (UI/UX, performance, security checks). Coordinate the public release with marketing and customer support readiness. *(Goal: launch confidently with a quality product and prepared support structure.)*

By time-boxing these phases, the team turned their roadmap into a deliberate delivery plan. It wasn't set in stone, they would review and adjust the plan every couple of Sprints, but it gave everyone a shared expectation of *what* would be released *when* and *why*. Instead of one overwhelming release, they had a sequence of smaller, learn-and-adapt launches. Team Vertex could now answer leadership's question with confidence: **"We plan to go live internally by Sprint 4, run a beta by Sprint 7, and launch to the public by Sprint 9."**

AGILE IN MOTION

Sprint 0 showcased several Agile and SDLC best practices in action. The team wasn't just planning; they were **living Agile values** from day one. Key practices included:

- **Stakeholder Mapping:** Rather than letting "everything be priority," the team proactively identified stakeholders and set engagement rules. This prevented feature creep driven by outside opinions and kept the Product Owner in the driver's seat for scope decisions. Clear stakeholder analysis ensured that expectations were managed and that the *right voices* were heard at the *right times,* aligning the project with business needs without derailing it.
- **Live Risk Board:** The team maintained an active risk log that was updated at each retrospective. This lightweight risk board (their "Risk Radar") made risk management a continuous, transparent process. By surfacing risks early and often, and attaching mitigation plans to each, Team Vertex embraced the Agile principle of adapting to change. They even tied risk management into backlog grooming (using **MoSCoW prioritization** to fight scope creep) so that potential issues like "MVP creep" or technical uncertainties were addressed before they could become fires.
- **Release Strategy:** Agile release planning turned their high-level roadmap into a phased rollout strategy. By dividing the release into **Alpha, Beta, and Public** phases, the team built in natural feedback loops and risk checkpoints. This practice reflects Agile principles of early and frequent delivery and welcome change. Internal users would validate the core, beta users would give usability feedback, and final adjustments could be made before the big public debut. The result was a transparent release plan that everyone, from developers to executives, could understand and track.

Each of these practices kept the team aligned, adaptable, and user-focused. Sprint 0 wasn't about writing exhaustive documents; it was about setting up frameworks (vision, scope, stakeholder engagement, risk approach, release cadence) that would guide the team through the Sprints to come.

TEAM REFLECTION

Sprint 0 had a profound effect on Team Vertex's cohesion and morale. At the start of the week, this was a group of individuals from different departments, unsure of how to begin. By the end, they had forged a common vision and way of working. Early uncertainty gave way to confidence as the team saw their ideas take shape in tangible plans. Key alignment moments, like the vision workshop where everyone articulated success in their own words, or the stakeholder session where they collectively said "no" to distractions, built trust and camaraderie.

They weren't just checking off planning documents; they were **building a shared mindset**. Maria's clear vision united them with a sense of purpose. Maya's facilitation of scope and risk made everyone feel heard and prepared. Jenna, Malik, Alex, and the rest contributed their expertise to shape the plan, seeing how their diverse skills fit together. By the end of SDLC phase 1 for Sprint 0's conclusion, Team Vertex felt *ownership* of the product and process. They moved forward not as a tentative new team, but as a confident, aligned unit with a clear sense of purpose. Sprint 0 proved that they were ready to tackle the real work together.

(As one team member put it, "Sprint 0 showed us we're not guessing: we know why we're here and how we'll win.")

Sprint 0: Exercises & Activities

Quick Check:

- How does having a clear product vision affect Sprint Planning and backlog prioritization?
- In your own words, why should a product vision be short, emotional, and user-centered?
- Which part of the Agile Project Scope do you think prevents the most misunderstandings during a Sprint, and why?
- If your team skipped defining scope in Sprint 0, which Agile value(s) would be most at risk?
- How might aligning stakeholder expectations early reduce rework or conflicts during Sprint execution?
- How could making risk identification a regular Sprint planning habit protect your team from future "fire drills"?

Apply This: *Exercise: Craft Your Agile Product Vision*

Instructions: You've just joined the team, kicking off a new PetTracker initiative.

Before jumping into user stories or backlogs, you've been asked to write and evaluate a clear product vision. Complete the steps below:

Step 1: Rewrite the Weak Vision

Here's a placeholder product vision someone proposed:

"Be the #1 solution in pet care tech using intelligent automation and insights."

Rewrite it using what you've learned. Make it **user-focused**, **problem-oriented**, and emotionally clear.

Your Improved Product Vision: ______________________________________

Step 2: Build Your Own Vision (Optional)

Imagine you're launching a simple app called **DogWalker** that helps professional dog walkers manage their routes, schedules, and payments. Based on that idea, draft a product vision for DogWalker.

Your Product Vision for DogWalker: ________________________________

Step 3: Vision Alignment Check

Use Maria's three key ingredients of a strong Agile vision to check your version. Does your vision clearly convey:

- **Why** does the product exist (the core purpose or problem it addresses)?
- **Who** it's for (the target user/audience)?
- **What** outcome or change does it aim to achieve (the impact on the user's life or work)?

Fill in the table below to ensure your vision covers all three elements:

Vision Element	✓ Present?
Why does the product exist? (Purpose)	[ ]
Who is it for? (Audience)	[ ]
What change or value does it create? (Impact)	[ ]

- **Apply This:** *Exercise: Linking Agile Principles to Project Scope*

Instructions: Below are key components from the PetTracker Agile Project Scope. Match each component to the relevant Scrum role, Agile value, or Scrum artifact/event that it supports. Then briefly explain why that component matters in Sprint 0.

Scope Component	Related Scrum Role / Event / Value	Why It Matters in Agile (Sprint 0)
1. Product Vision		
2. Definition of Done		
3. Working Agreements		
4. MVP vs. MMP Breakdown		
5. Core Team Roles & Skills		
6. Success Criteria		
7. Stakeholder Identification		
8. Schedule (1-week Sprint 0 + 4 Sprints)		

(Reflect: How do each of these scope components tie back to Agile/Scrum fundamentals, and what would happen if any were overlooked during planning?)

- **Role Replay:** *Exercise: Map Your Stakeholders Before They Map Your Scope*
  Scenario: You're helping Team Vertex finalize their stakeholder strategy for the PetTracker MVP. Multiple voices want input, but not all stakeholders need the same level of engagement. Your task: identify key players, map their influence, and plan how and when to involve them.

Part 1: Mission Champion Challenge

Imagine that your key champion stakeholders suddenly goes on leave for a month. How would you keep the momentum going and ensure their input is still considered?

Part 2: The Compliance Curveball

Suppose a new stakeholder from the compliance department is introduced halfway through the project, qnd they have a lot of concerns that could potentially slow things down. How would you integrate them into the project without derailing your timeline.

Part 3: Reflection (consider this after mapping). How might aligning stakeholder expectations early reduce rework or conflicts during development? (Refer back to your map and engagement plans for clues.)

- **Role Replay:** *Exercise: Surface Risks Before They Surface You*
  Scenario: You're working with Team Vertex during the Planning/Initiation phase of PetTracker. The team just drew a big "Risk Radar" on the board. Your job is to anticipate potential risks for the upcoming release and decide how the team should respond.

Part 1: Identify 3 Risks

List three realistic risks that could impact the team's ability to deliver a feature (for example, the *Walk Logging* feature or the *Medication Reminder* feature). For each risk, specify the type and give a short description.

Risk Type (Tech, Delivery, etc.)	Description of Risk	Feature Affected

Part 2: Choose a Risk Response

For each risk above, decide how the team should respond (Avoid, Mitigate, Transfer, or Accept) and briefly describe the plan:

Risk (from Part 1)	Chosen Response (A/M/T/Acc)	Your Plan (summary)

Part 3: Plot One Risk on the Radar

Pick one of your listed risks and consider its likelihood and impact. Mark where it falls:

Likelihood: ☐ Low ☐ Medium ☐ High
Impact: ☐ Low ☐ Medium ☐ High
Risk Description: ____
Why it matters: ____

(Think: By regularly discussing risks like this each Sprint, how is the team better protected from last-minute fire drills?)

- **Role Replay:** *Exercise: Plan the Agile Release like Team Vertex*
  Scenario: Your PetTracker MVP is nearly ready. Now it's time to move from "done" to "delivered." Work alongside Team Vertex to sketch a phased release plan that supports feedback, flexibility, and flow.

Part 1: Release Confidence Check

Now that you're planning your Agile release, let's think about what will give you confidence that each feature is truly ready to go live. For this exercise, outline at least three "confidence criteria" that would signal a feature is ready for release. For example, you might decide that a feature is ready if it passes all user acceptance tests, receives positive feedback from a pilot user group, or meets certain performance benchmarks.

By defining these confidence checks, you're creating a clear set of criteria that will help your team know when it's truly time to hit the "release" button.

Part 2: Build the Success Criteria (Beta Phase)

For the Friends & Family **Beta** phase, select **two** success metrics you would prioritize to gauge its success:

☐ > 90% positive usability feedback

☐ At least 3 high-priority internal bugs resolved
☐ Load-tested to 10,000 concurrent users
☐ ≥ 10 beta users actively complete core flows daily
☐ API average response time < 300 ms under load
☐ Support team trained on new features and ready for launch
(✓ Select two that best indicate a successful Beta)

Part 3: Spot the Planning Traps

Which of the following planning choices would **go against Agile release principles**? (Select all that apply)

☐ Releasing all features to all users at once
☐ Building in buffer time between release phases
☐ Involving the marketing team before go-live
☐ Using feedback from internal alpha users to improve the product
☐ Locking the full scope and dates for all releases before Sprint 3 starts

Part 4: Define "Launch Ready"

You're finalizing the Public Launch checklist. Which of the following should be included as **launch readiness** items? (Select all that apply)

☐ UI/UX design polish complete
☐ Final end-to-end QA testing passed
☐ Key stakeholder sign-off obtained
☐ Push notification content written and tested
☐ All developers scheduled a vacation immediately after launch

(Think about why each item is or isn't essential for a smooth public release.)

- **Your Toolkit:** *Reference Tables & Templates from Sprint 0*

Vision vs. Goals vs. Roadmap: To avoid confusion between big-picture vision and shorter-term targets, the team clarified the difference:

Element	Purpose	Time Horizon	Example (PetTracker)
Vision	Why we exist	Long-term (years)	Make pet care easier for working families."
Goal	What we want to achieve	Mid-term (quarterly)	"Reduce missed pet feedings by 30%."

Roadmap	How we plan to get there	Short-term (Sprints)	"Build logging feature, add reminders, summarize routines."

Good vs. Bad Product Vision Examples: Use these examples as guidance when crafting or evaluating a vision statement (see table in the *Team Vertex in Action* section above for details on *why* each vision succeeds or fails).

MVP Scope vs. MMP (Planned Later): Early in Sprint 0, the team defined the initial feature scope for the MVP (first 4 Sprints) and set aside other features for the MMP (next phase) to maintain focus:

MVP (Sprint 1–4): Core Features:

Feature	Description
Profile, Login & Auth	Email/password signup, secure session handling; add pet details (name, breed, age, photo).
Log Feeding	Record meal details (food type and portion size).
Walk Log	Track walks with a timer and background GPS route.
Medication Log	Log medications with name, dosage, time given, notes.

Out of Scope for MVP (Planned for MMP):

Deferred Feature	Reason for Deferring (to MMP)
Multi-pet support	Adds complexity to data model (can wait).
Vet visit history	Needs secure data storage (compliance heavy).
Real-time route tracking	Performance overhead on mobile devices.

(By keeping the MVP scope intentionally tight and deferring nice-to-haves, the team left room for learning and iteration.)

Stakeholder Map (Roles & Engagement): An initial Stakeholder Map outlined who's involved and how:

Stakeholder	Role	Engagement
Product Sponsor	Budget & release sign-off	Weekly review
Vet Consultant	Domain expert (pet health)	Ad-hoc input

QA Lead (Malik)	Quality strategy owner	Quality strategy owner
Design Lead (Jenna)	UX/UI ownership	Sprint-level involvement
End-User Proxy	Represents pet owners' needs	Via UX research (ongoing)

(This map ensured each role knew how they'd contribute and when: it also prevented surprises by involving experts like the Vet Consultant or Compliance at the right time.)

Engagement Plan: When planning stakeholder involvement, Team Vertex followed these guidelines:
- Not everyone needs to attend every Sprint meeting. Tailor engagement by stakeholder relevance and influence.
- Define **cadences**: e.g., Executive sponsor gets quarterly roadmap reviews; End users give input via beta feedback; Compliance gets early access to features affecting regulations.
- Co-create priorities with stakeholders, but maintain the Product Owner's authority to say *"not now"*.
- Use visual aids (stakeholder maps, RACI charts) to keep everyone's roles and communication channels clear.

Agile Live Risk Board (Sprint 0 Snapshot): A glimpse at Team Vertex's risk board with some risks and mitigation plans identified in Sprint 0:

Risk Area	Mitigation Plan
API delivery delays	Use mock API endpoints for early development.
GPS cross-platform issues	Time-box a spike in Sprint 0 + do device compatibility testing. Use mock API endpoints GPS cross-platform issues for early development.
Data duplication bug	Establish data validation patterns + have QA focus on edge cases.
"MVP creep" (scope growth)	Apply MoSCoW prioritization + regular GPS cross-platform issues backlog grooming to cut nice-to-haves.

(The risk board is revisited each Sprint, new risks added and resolved ones retired, so the team is never blindsided.)

Agile Risk Categories: Common risk types and PetTracker examples the team identified:

Risk Type	Description	Example (PetTracker)
Technical	Code, architecture, or integration challenges	Third-party GPS API changes without notice.
Delivery	Timeline, capacity, or overcommitment issues	Jenna's PTO overlaps a critical design Sprint (capacity drop).
Requirements	Shifting scope or unclear requirements	VP adds a new feature mid-Sprint (scope change).
Dependencies	External or shared resource constraints	Backend dev is split with another team (resource bottleneck).
Compliance	Regulatory, legal, or policy concerns	Pet data storage must meet privacy rules (GDPR/HIPAA related).

Risk Response Strategies: The four ways to address risks, as discussed by Maya and Malik:

Strategy	Description	PetTracker Example
Avoid	Change the plan to eliminate the risk	Use an in-house geo-location module instead of a risky third-party API.
Mitigate	Reduce the likelihood or impact of the risk	Write early integration tests to catch API changes quickly.
Transfer	Shift the risk to a third party or external entity	Ask the API vendor for an uptime guarantee (SLA) or involve Legal for compliance risk.

Accept	Acknowledge the risk and live with it (monitor)	App Store review delays. Understand it's beyond control and plan around it.

Agile Project Scope (PetTracker): Sprint 0 Summary

(This one-page Agile Scope document captures all key planning elements the team agreed on during Sprint 0.):

Section	PetTracker Agile Scope Details
Product Vision	Empower pet owners to care for their pets through smart, real-time monitoring and daily health tracking.
Objective Statement	Deliver an MVP mobile app that allows users to log and manage feeding, walking, and medication routines, ensuring no pet care activity is missed.
Deliverables	MVP: Meal logging, walk tracking, medication reminders, basic pet profile setup. Next (MMP): Vet appointment scheduling, smart collar integration, lost pet alerts, community features.
Stakeholders	Executive Sponsor; Customer Advisory Panel; Legal/Compliance Team; Technical Architect; Marketing; Vet Consultant; Pet Parent (end user) representatives.
Core Roles & Skills	Maya (Scrum Master); Maria (Product Owner); Alex (Frontend Dev); Malik (QA Lead); Jenna (UX/UI Designer); You (Backend/API Dev). (Cross-functional team with all skills to deliver increments.)
Schedule & Cadence	1-week Sprint 0 for setup/planning, followed by 4× two-week development Sprints. Target internal MVP release at end of Sprint 4.
Working Agreements	Daily stand-ups @ 10 AM EST; use Slack for async updates; Figma for design reviews; Sprint Review every second Friday with stakeholders.
Definition of Ready	User story has clear acceptance criteria, UX design attached, APIs identified, dependencies flagged, and story sized/approved by the team.
Definition of Done	Code is merged to main branch, peer-reviewed, all tests passed, feature demoed to PO and accepted, meets DoD checklist (incl. regression test, updated documentation).
Success Criteria	MVP usage and quality metrics: App successfully launches on app store; < 5% crash/bug rate in production; 90%+ user retention after 30 days; positive feedback from beta users.
Communication Plan	Weekly stakeholder sync on Fridays; daily Slack summary of progress; Confluence pages for all important decisions and project artifacts (updated regularly).

Tools & Requirements	Tools: Jira (backlog), GitHub (repo), Figma (design), Postman (API testing). Key Requirements/Standards: Ensure data encryption (pets + owner info privacy), iOS/Android cross-compatibility, and offline functionality for core features.

(This Agile Scope acts as the team's charter, stored in Confluence and treated as a living document to be revisited if anything major changes.)

Stakeholder Engagement Matrix (Sprint 0 Example): During Stakeholder Mapping, Team Vertex created a matrix for key stakeholders to align on influence and engagement:

Stakeholder	Role/Interest	Level of Influence	Engagement Strategy
VP of Product	Project Sponsor (Champion)	High	Weekly check-in meetings; roadmap reviews every Sprint
Customer Support Lead	Represents user feedback	Medium	Invite to Sprint Reviews for direct feedback from a support perspective.
Lead Vet Consultant	Pet health domain expert	Medium	Include in backlog grooming or story mapping for medical features.
Compliance Officer	Governance & data privacy	High	Early involvement in UAT for any feature handling personal data.
CTO	Technical oversight	Medium	Monthly architecture demos; ensure alignment on tech decisions.
Pet Owner Advisory Group	End-user representatives	Low	Quarterly surveys; include in Friends & Family Beta group for testing.

(By categorizing stakeholders this way, the team ensured high-influence stakeholders were closely managed and low-influence ones were still heard appropriately. This minimized last-minute surprises like "Where's that feature I wanted?" from a sponsor.)

Lightweight Risk Management Tools: The team considered several Agile-friendly tools to keep risk management practical and ongoing:

- *Risk Radar Chart:* Plot each risk on a chart by *Likelihood* (x-axis) and *Impact* (y-axis) to visualize which ones are critical (high/high) vs. minor.
- *3-Column Risk Wall:* Simple board with columns "What could go wrong?", "How likely is it?", and "What will we do?" It forces the team not just to list risks but also to assess and plan responses.
- *Story Risk Tags:* Mark individual user stories with risk flags or notes (e.g., "⚠ requires external API") so the team addresses risks during development, not

just in abstract.

- *Sprint Risk Retros:* In each retrospective, explicitly discuss: *"What nearly went wrong this Sprint? What did we learn? What will we watch out for next time?"* This normalizes talking about risks regularly.

Release Phase Definitions: When planning their release strategy, Team Vertex agreed on what each phase means:

- **Internal Alpha:** A private release to team members or internal staff, aimed at validating core functionality and stability in a controlled environment. (Think of it as "testing in our own backyard" before any outsiders use the product.)
- **Friends & Family Beta:** A limited release to a small, trusted external group (friends, family, or select friendly customers). The goal here is to gather usability feedback and catch bugs in a real-world context, while the audience is still contained.
- **Public Launch:** The full release to all users. By this stage, performance is optimized, all major bugs are resolved, and support & marketing are prepared for broad adoption. This is the official "go live" moment, ideally with confidence built from the alpha and beta phases.

Agile Release Plan: Phased Launch Roadmap (Team Vertex)

(This table shows how Team Vertex mapped out their MVP release across phases and Sprints.):

Phase	Goal of Phase	Scope Highlights	Target Date
Internal Alpha	Validate core stability	• Core logging features (meals, walks, meds) • (Internal testers only)	End of Sprint 4

Friends & Family Beta	Collect early user feedback	• Basic in-app reminders • Daily Summary Dashboard • Vet-visit scheduling • Fixes from Alpha testing	End of Sprint 7
Public Launch	Launch with confidence	• Data export/sharing feature • Referral program integration • Full UI/UX polish • Final performance tweaks	End of Sprint 9

Details: By **Sprint 4**, the plan was to have an MVP internally tested and approved by stakeholders (internal alpha). By **Sprint 7**, additional features would be completed and a beta released to a small external group, incorporating feedback from the alpha. By **Sprint 9**, all remaining features and polish would be done for the public launch. This phased approach allowed continuous feedback and course-correction, dramatically reducing risk for the final release.

(For each phase, the team also outlined "exit criteria." E.g., for Alpha: all critical bugs fixed; for Beta: ≥ 10 beta users actively using the app with positive feedback; for Public: no high-severity bugs open, etc. These criteria were part of the team's definition of "ready to move to next phase.")

CHAPTER SUMMARY

In Sprint 0 (SDLC Phase 1: Planning/Initiation), Team Vertex transformed from a newly assembled group into a focused, aligned team with a game plan. They established a compelling **product vision** that became their guiding star and defined the **scope** of their first release, clearly distinguishing MVP must-haves from later nice-to-haves. They achieved early **stakeholder alignment** by mapping out who needed to be involved and how, preventing miscommunications and scope creep. The team identified key **risks** upfront and decided how to handle them, instilling confidence that they could tackle the unexpected. Finally, they sketched an Agile **release strategy** to deliver value in incremental phases (internal alpha, beta, public launch) rather than a risky big bang. By the end of Sprint 0, Team Vertex had all the foundational elements in place: vision clarity, an agreed scope, stakeholder buy-in, a risk mitigation plan, and a phased release roadmap. No working software had been shipped yet, but the team's groundwork ensured that when Sprint 1 begins, they will move forward with **confidence, alignment, and a clear sense of purpose**.

SDLC PHASE 1: PLANNING/INITIATION/SPRINT 0

Laying the groundwork for everything to come together.

PRODUCT VISION

Anchor the team with a clear, user-focused outcome.

PROJECT SCOPE

Define roles, norms, and team agreements.

STAKEHOLDER ENGAGEMENT

Identify who matters and how to engage them.

RISK MANAGEMENT

Spot early risks and plan lightweight responses.

RELEASE PLANNING

Phase the rollout (Alpha, Beta, Public) and buffer time for testing, bugs and feedback fixes.

SPRINT 0

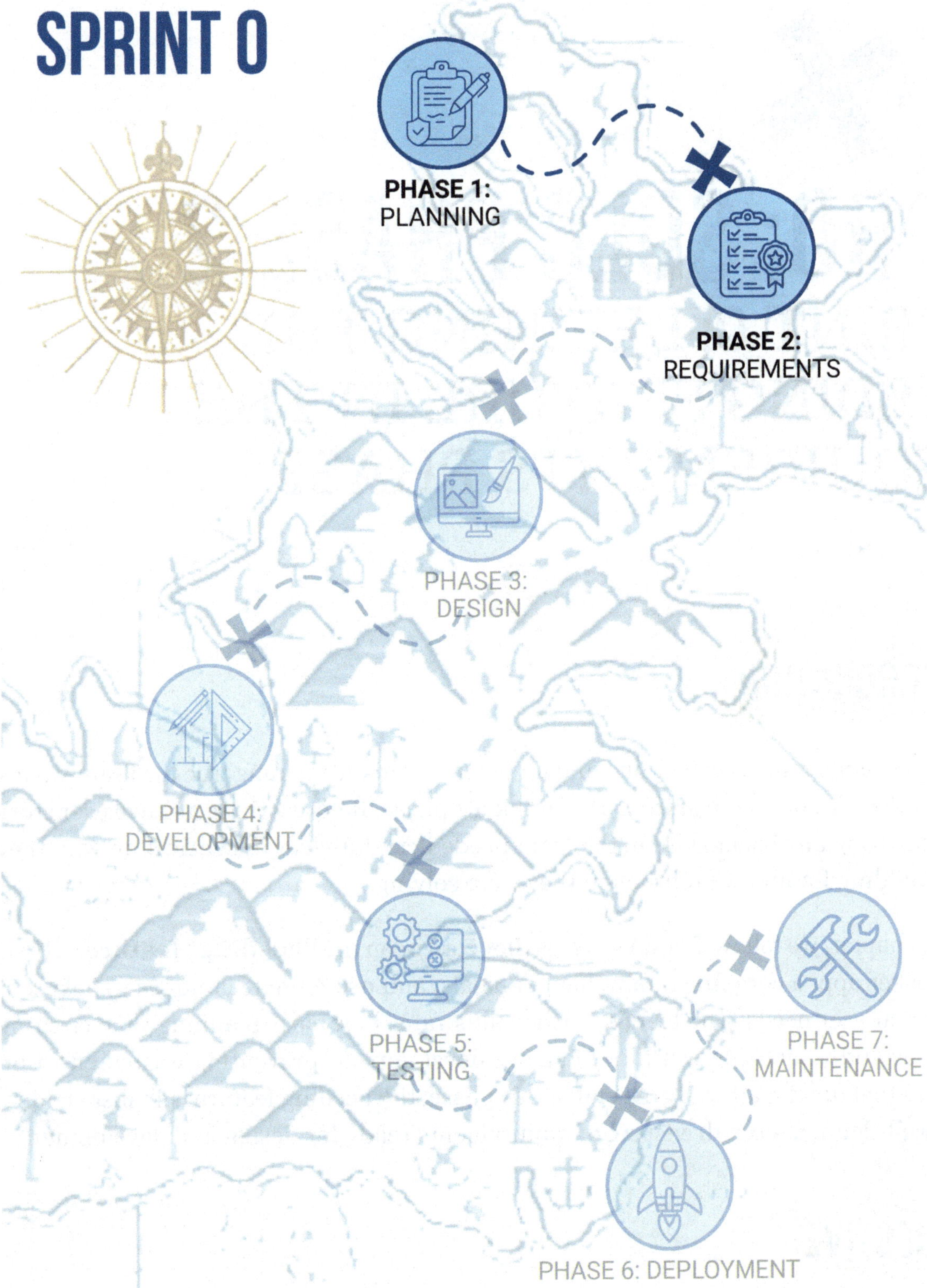
PHASE 1:
PLANNING
PHASE 2:
REQUIREMENTS
PHASE 3:
DESIGN
PHASE 4:
DEVELOPMENT
PHASE 5:
TESTING
PHASE 6: DEPLOYMENT
PHASE 7:
MAINTENANCE

9.

SPRO: SDLC PHASE 2

SPRINT 0: DEFINING THE FEEDING FEATURE IN THE REQUIREMENTS PHASE

INTRODUCTION

"If you build for everyone, you build for no one." This adage rang true for Team Vertex at the start of their **Sprint 0** requirements phase. Before writing a single user story or prioritizing any backlog item, the team needed crystal-clear **direction** on who they were building for and what problem they were solving.

In this chapter, we follow Team Vertex as they zero in on the **"Feeding" feature** of their PetTracker app, a core function to log pet meals, and use Agile techniques to ensure it meets real user needs. The team's journey illustrates how early planning (like creating personas and visualizing workflows) guides development, preventing wasted effort on features that don't matter. By the end of this phase, the Feeding feature isn't just an idea; it's a well-defined, user-driven set of requirements ready for design and development.

OBJECTIVE

Sprint 0: Requirements Phase teaches the team how to **gather and clarify requirements** before any coding begins. Key learnings include: conducting stakeholder workshops, JAD session or interviews to understand user needs, using **visual modeling**

(use case diagrams, workflow charts, activity diagrams) to map out processes, facilitating **story mapping** sessions to translate workflows into user stories, and applying prioritization methods (like **MoSCoW**) to decide what's essential for the first release. In short, this phase is about turning fuzzy ideas into a concrete plan. By focusing on a specific feature (Feeding) as an example, the team learns how to identify user expectations, capture edge cases, and build a **prioritized backlog** that will guide development in upcoming Sprints.

FEATURE IN FOCUS: FEEDING

The team gathers for a tight retrospective on testing capability: Did testing slow anyonedown? Can QA work without Dev breathing down their neck? Are we catching risk long

before failure reaches users? challenge. Emily is a busy dog mom who sometimes gets caught up in work and forgets if or when she has fed Luna, leaving the pup waiting and Emily feeling guilty. The Feeding feature of PetTracker is designed to solve this problem. It allows users to log each meal (what was fed and when) and aims to ensure no meal is ever missed or double-fed. By tracking feedings, the app can spot patterns (for example, if lunchtime feedings are often late) and eventually send timely reminders.

The Feeding feature thus accomplishes two things for users: it provides peace of mind that their pet's diet is properly recorded, and it offers insights into the pet's eating habits over time. During Sprint 0, Team Vertex identified feeding as a must-have capability, one that would deliver immediate value by addressing a primary pet care need. In fact, they elevated "Meal Logging & Tracking" as one of the central feature for the product, ensuring that features like "Log a Meal" and "View Feeding History" would be prioritized for the Minimum Viable Product (MVP).

TEAM VERTEX IN ACTION

To bring the Feeding feature from idea to reality, **Team Vertex** (Maria: Product Owner, Maya: Scrum Master, Alex: Developer, Jenna: UX Designer, Malik: QA, and *You*: an Agile team member) rolled up their sleeves and went through a series of collaborative steps:

- **Defining Roles & Personas:** Before deciding *what* to build, the team clarified *who* they were building for. Maria organized a workshop to develop user personas from stakeholder interviews. They created three core personas to represent their target users: **Pet Parent Emily** (a tech-savvy owner of a Labrador, diligently tracking health data), **Busy Parent Ben** (juggling multiple pets, needing quick logs and reminders), and **Veterinarian Vicky** (reviewing shared logs for accuracy).

- These personas gave life to the generic user roles the app would serve (Pet Owner as the primary user, with potential Guest and Admin roles for future consideration). Emily's persona, in particular, put a face to the Feeding feature. Her pain points of forgetting meals and her goal of logging care in under 2 minutes became guiding insights. When Maria shared Emily's story with the team, it immediately clicked that **reliable meal logging and notifications** would be critical. The persona exercise ensured everyone on Team Vertex kept the real end-user in mind from the start.
- **Use Case Diagram, Emily's Actions:** With personas established, Jenna sketched a high-level **use case diagram** to outline what a Pet Parent like Emily needs to do in the app. This diagram mapped **Actors** (e.g., Pet Parent, Secondary Caregiver) to **Use Cases** (key actions):
  - Log Feeding
  - Log Walk
  - Log Medication
  - View Daily Summary
  - Set Reminders
  - Share Care Summary

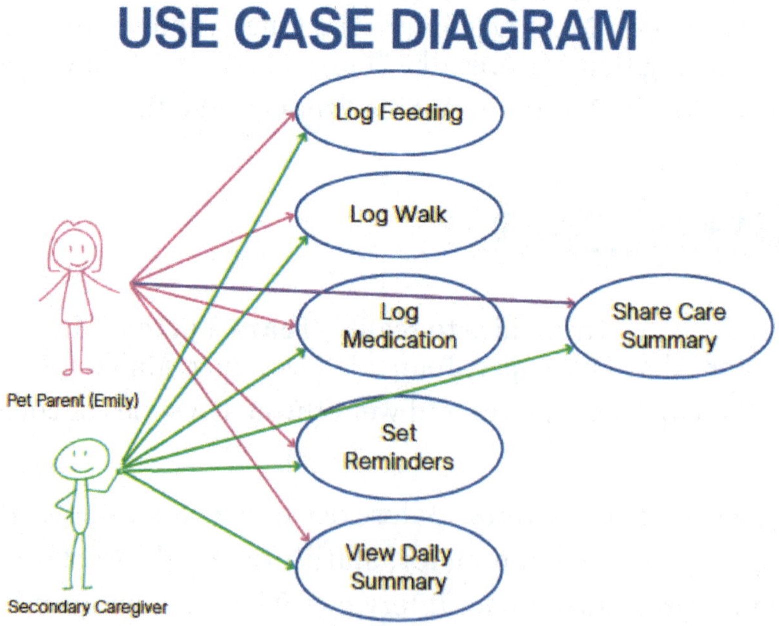

By visualizing all possible interactions, the team could see the full scope of functionality at a glance. For example, the use case diagram highlighted that *Emily can log a meal and view summaries*, whereas her roommate (a secondary caregiver persona) might only have viewing permissions. This bird's-eye view helped spot any missing interactions early and gave context to where the Feeding feature sat among other features. It was clear that "Log Feeding" was a central use case for the primary Pet Parent user, a foundational piece of the app's value.

- **High Level Workflow Diagram:** Instead of diving straight into a single feature like feeding, the team is assembling a holistic user journey map. Alex and Jenna are sketching out the complete flow—landing on the home page, registering or logging in, creating a pet profile, then logging meals, walks and medication—to understand how each major touchpoint connects. This high level view ensures everyone aligns on the overarching experience and exposes gaps or dead ends before they build individual features. In text form, the flow looks like: Start → home page → registration/login → create pet profile → log/edit/save meals → log/edit/save walks → log/edit/save medication → End.

HIGH-LEVEL PET CARE WORKFLOW DIAGRAM

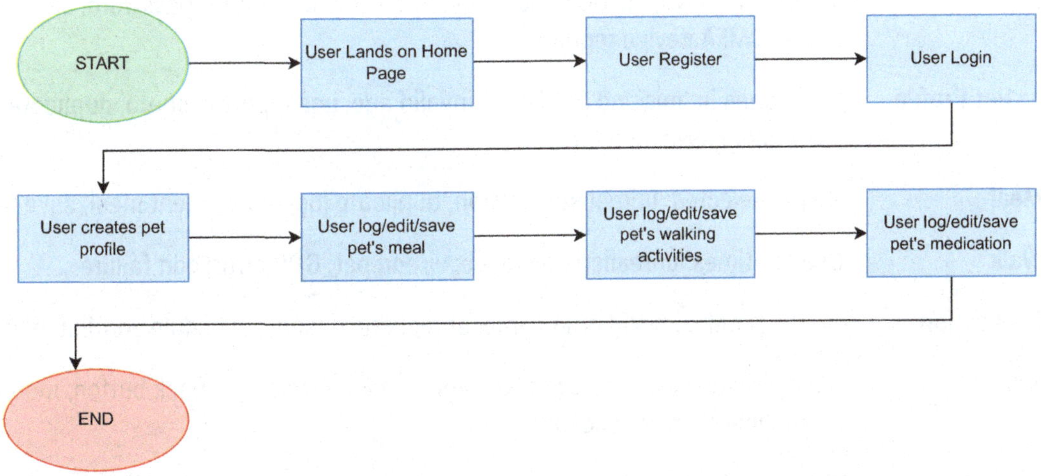

- **Activity Diagram, Edge Cases and Conditions:** Having mapped the basic flow, the team dug deeper by creating an **activity diagram** for the Feeding feature. Maya led the effort to chart out all the decision points and alternate paths involved in filling out the feeding form. This more detailed diagram considered questions like: What if the user doesn't select a pet? What if they try to set a feeding time in the future, or leave required fields blank?

> The activity diagram for "Fill Feeding Form" included conditional branches for these scenarios. For example, if no pet is selected, the system should display a "Please choose a pet" error and loop back so the user can fix it.

Similarly, if the chosen time is in the future, it should warn "Cannot log future time" and prevent submission. The diagram also showed parallel actions, for instance, when the user taps "Submit," a backend API call is made to save the data while a loading spinner animates on the UI in parallel. By visualizing the **validations and parallel processes**, Team Vertex identified potential *"edge cases"* (like a server error if the database is down, or the app being offline) and noted where they might need to add technical research spikes or extra acceptance criteria. This exercise assured that the Feeding feature's design was robust, covering not only the sunny-day scenario but also what happens when things go wrong (e.g., if the server returns an error, the app should show "Unable to save. Try again.").

Activity	Potential Edge Cases
Home/Landing	Page refresh/close; restricted feature without login; network outage
Registration	Invalid email; duplicate email; password mismatch; missing fields; decline terms; rate limit; expired verification
Login	Wrong password; account lock; unverified account; forgot password; session expired; MFA device required
Create Pet Profile	Not logged in; missing pet fields; invalid age; unsupported photo; duplicate name; incomplete save; timeout
Log Meal	No pet selected; unrealistic portion; duplicate log; nonexistent meal; save failure
Log Walk	Overlap times; unrealistic distance; wrong pet; GPS error; edit failure
Log Medication	No pet profile; invalid drug name; unrealistic dosage; schedule conflict; edit error
General	Connectivity loss; concurrent sessions; resubmission via back button; mobile app killed; unauthorized access

Every path in the activity diagram turns into a concrete backlog item. Some paths become user stories, which describe behavior the system must support. Other paths reveal unknowns that need investigation first.

In Agile, those investigation tasks are called **spikes** — short, time-boxed research efforts used to reduce uncertainty before committing to a solution.

For example, the check "If time > current time" became a clear rule in the story's acceptance criteria. Meanwhile, questions about how the app should handle offline submission were captured as a technical spike to explore options and constraints before development begins.

- **From Diagrams to User Stories:** Once the workflows and activities were mapped out, Maria and Alex began extracting concrete backlog items. Each step or decision from the diagrams turned into a rough **user story or spike**.
  - For instance, the action *"Display 'Log Feeding' screen"* became a user story: *"As a pet parent, I want to see a Log Feeding form with pet selection, meal type, time picker, and notes fields."*
  - Likewise, a branch like *"Check GPS permission"* (relevant for walk logging) might become a Spike: *"Research GPS permission handling on iOS/ Android."*.

By doing this, the team ensured **traceability**; every piece of the Feeding workflow was accounted for in the backlog. Nothing would slip through the cracks during development because the diagrams had fleshed out all the actions (stories) and questions (spikes). This approach also made writing acceptance criteria easier later on, since the team had clearly defined what should happen at each step (e.g., *"If no pet is selected, display an error message and do not save the entry"* corresponds to a test condition).

- **Story Mapping Workshop:** With a solid grasp on the feeding process, Team Vertex shifted to **User Story Mapping** to see the bigger picture. In a whiteboard session led by Maya, they mapped out the *entire user journey* of the PetTracker MVP with the assistance and feedback from the Vet Consultant, and they positioned the Feeding feature within it. They identified the top-level user activities a pet parent goes through in a typical day.

The backbone of their story map had three major activities (swimlanes) drawn from Emily's routine: **Feed Luna**, **Walk Luna**, and **Give Medication**. Under each of these, the team listed specific user tasks as stories, organizing them by priority from top (must-haves) to bottom (nice-to-haves). For **Feed Luna**, the must-have stories they wrote included:

- ✅ **Must:** *"As a pet parent, I want to log a meal so I can track Luna's feeding consistency."*

- ✓ **Must:** *"As a pet parent, I want to view a list of past meals so I can see Luna's feeding history over time."*
- **+ Could:** *"As a pet parent, I want to select a meal type (e.g., Breakfast, Lunch, Dinner) so that entries are consistent."* (a nice-to-have, since the app could technically function without categorizing meal types at first)

They performed the same exercise for the **Walk Luna** and **Give Medication** activities, resulting in their initial universe of user stories (e.g., logging walk events, viewing walk history, logging medications, etc., each with Must/Could designations). Once all the stories were mapped, Team Vertex drew a **horizontal line** across the map to slice out the MVP. Everything above the line would be included in the first release, and everything below could wait until later.

For Feed Luna, this meant the two "Must" stories (log a meal, view meal history) were above the line, and the "Could" story (meal type tagging) was below.

"This lets us release something valuable fast without being stuck waiting for bells and whistles," Maria noted, as they all agreed only the essentials should ship first.

The story map was an eye-opener for the team. It ensured that features like Feeding were not considered in isolation but as part of a **cohesive user workflow**. Jenna commented that mapping the stories helped her envision the design flow across screens rather than designing isolated pages. Malik realized he could derive more scenario-based test cases (like checking the continuity from feeding to walking flows). In short, story mapping helped the team see the app *"from the user's eyes, not just the backlog's,"* as Maria put it.

After story mapping, the team was left with a wall of sticky notes: dozens of granular user stories. To bring structure to this potential chaos, they began clustering related stories into **Epics**. Epics are larger buckets of work that group stories by a shared theme or goal.

Before they could define their Epics, the team aligned on a **Theme**.

Theme: *A Theme is a high-level strategic objective that connects related Epics under a common goal. It provides direction by explaining* ***why*** *the work matters, while Epics break that work into major capabilities that deliver user value.*

Standing in front of the story map, Maria guided the discussion: *"These stories aren't isolated; they cluster around broader care routines."*

Through this process, Team Vertex organized its backlog into one unifying Theme and four foundational Epics:

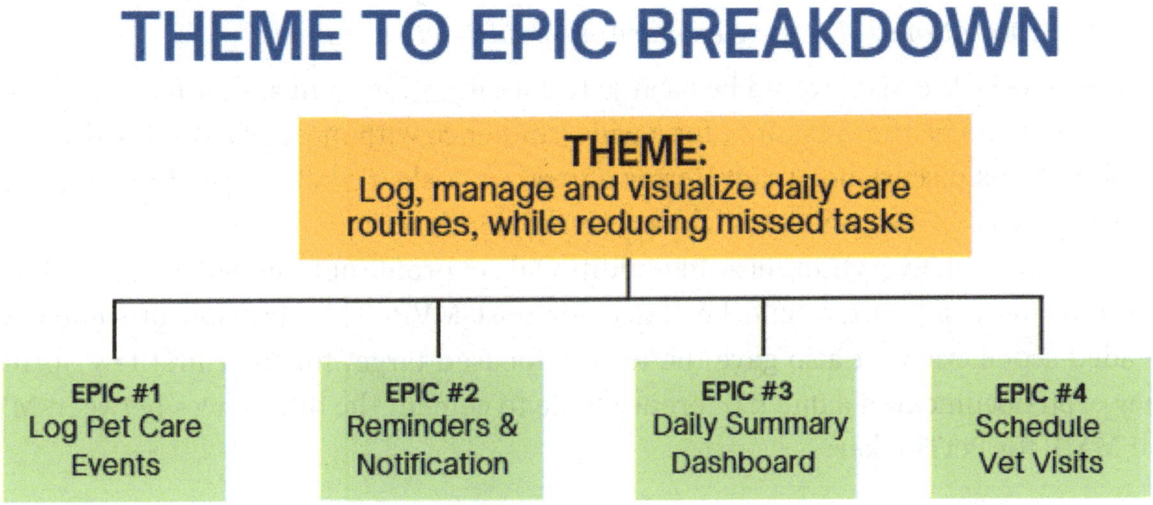

Each Epic represented a major capability that delivered distinct user value in the PetTracker app. More than just categorization, this grouping gave the team a strategic framework. By sequencing work logically (e.g., building the core logging epic before layering on dashboards and scheduling), they created a roadmap that was easier to communicate with stakeholders and to manage at a high level.

- **MVP vs. MMP, Deciding What's Essential:** With epics in hand, the team confronted a critical product question: **What features are absolutely needed for launch, and what can wait?** In Agile terms, what is the *Minimum Viable Product* vs. a more complete *Minimum Marketable Product*?

"We have a solid map of stories and epics, but let's be clear: not everything here will ship in the first version," she asserted during a Sprint 0 review meeting.

The team agreed to define their **MVP** strictly as the features that would validate the app's core value: allowing pet owners to log essential pet care events (meals, walks, medications) and get basic reminders. Fancy extras would be slated for the **MMP** (a more polished version once the concept was proven). In discussing MVP vs MMP, they collectively decided that: **logging** features are the non-negotiable core for MVP, whereas things like rich analytics, social sharing, and aesthetic enhancements are nice but not necessary at launch. For example, being able to log a meal or a walk was deemed a "Must Have" for MVP, but viewing a detailed daily trend chart could be an MMP item. Maya summed it up well: *"We haven't written a line of code yet; this is the time to set realistic expectations. MVP gets us learning. MMP gets us launching."*

In other words, the MVP would be used to test assumptions and gather feedback, while the MMP would be the version fit for a wider audience with more bells and whistles. By the end of this discussion, Team Vertex agreed on a clear MVP scope: **Log Pet Care Events** (meals, walks, meds) plus basic reminder functionality would constitute the MVP's feature set. Everything else, like additional pet profile options, advanced analytics, and multi-user support, would be slated for post-MVP. This decision prevented the dreaded scope creep. It also gave the team a focused target for their first few Sprints. They even documented a quick reference table to cement the differences between MVP and MMP for PetTracker:

Aspect	MVP (Sprint 0–7 focus)	MMP (Post-MVP)
Purpose	Validate core value quickly, learn fast	Launch confidently, attract real users
Audience	Early adopters, internal testers	Public app users, external stakeholders
Scope	Core functionality only (log meals, walks, meds; basic reminders): no polish or extras	Complete experience (reminders, summaries, dashboards, etc.) with full polish
Risk Level	Higher risk, fast feedback (we're testing assumptions)	Lower risk, high usability (refined based on feedback)
Example	Log meals only (core need addressed)	Log meals + set feeding reminders + view dashboard (extended features)

This clarity on MVP vs MMP guided all subsequent decisions. For instance, when someone floated an idea about a **social sharing** feature, it was easy to label it "Won't Have for MVP" and defer it. Team Vertex could now communicate to stakeholders exactly what the **first release** would (and would not) include. Maria felt confident defending these choices, knowing they were backed by user-driven logic. The whole team internalized the mantra that *"just enough is not the same as barely usable."* The MVP would do a few things **well**, rather than doing everything poorly.

Maya reminded everyone with a smile: "You don't win by launching everything. You win by launching the right thing, at the right time, for the right audience."

Initial Backlog Creation: Having nailed down the scope, Team Vertex proceeded to build the **Initial Product Backlog**. This was the tangible list of epics, features, and user stories that would drive Sprint planning. Maria kicked off a backlog brainstorming session by revisiting the product vision:

> **"EMPOWER PET PARENTS TO LOG, MANAGE, AND VISUALIZE DAILY CARE ROUTINES,**
> **WHILE REDUCING MISSED TASKS LIKE MEDICATION OR VET VISITS," SHE REITERATED,**

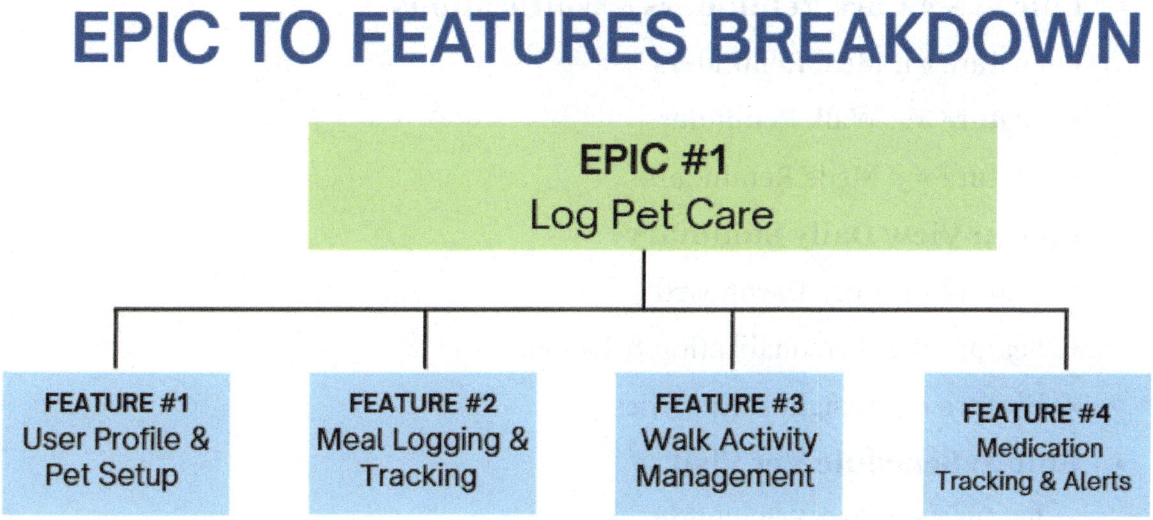

Using that vision as a compass, the team revisited their epics to ensure each was aligned with those goals. They then **decomposed each epic into 4 key features:** chunks of functionality small enough to build and test within a single Sprint. For example, the **Log Pet Care** epic (Epic 1) was broken down into these features:

The team performed similar high-level breakdowns for the Epic #1 (Profile Setup, Meal Logging, Walking Activity and Medication Tracking), ensuring each feature was **independently testable** and delivered incremental user value. Throughout this process, they discussed and noted any obvious **dependencies** or sequencing needs. For instance, to implement "Set Feeding Reminder," they'd first need the basic "Log a Meal" in place (no point reminding something that can't be logged yet).

Alex and Malik also put on their engineering hats to **estimate effort** and flag risks: they gave rough sizing (in story points) for each feature and identified if any feature needed a preparatory spike or design review. One by one, features were sticky-noted under their respective epics on the board.

By the end of this "Epic Decomposition" session, the team had a board full of epics with feature lists under each. It looked something like this:

- **Epic 1: Log Pet Care Events**
  - Feature #1: Profile Creation & Login
  - Feature #2: Log Meal
  - Feature #3: Log Walk
  - Feature #4: Log Medication
- **Epic 2: Set Care Reminders & Notifications**
  - Feature #1: Meal Reminders
  - Feature #2: Walk Reminders
  - Feature #3: Meds Reminders
- **Epic 3: View Daily Summary**
  - Feature #1: Core Dashboard
  - Feature # 2: Personalization & Layout
  - Feature #3: Insight & Analytics
- **Epic 4: Schedule Vet Visit**
  - Feature #1: Core Scheduling
  - Feature #2: Pre-Visit
  - Feature #3: Day-of Visit
  - Feature #4: Post-Visit & Follow-ups
  - Feature #5: Notifications & Calendar

This structured backlog was the bridge between the team's ideas and actionable work. Everyone could now see **exactly what to build**, and roughly in what order. Importantly, not everything in this list would be built at once, so the final step was to decide **what to tackle first** when Sprint 1 kicks off.

- **Prioritization; MoSCoW and More:** With the backlog features laid out, Team Vertex confronted the question of *priority*. What would they build in the very first Sprint versus later on? Maria had dozens of feature requests pinned up, each with its own champion or stakeholder backing. To cut through the noise, Maya introduced a simple prioritization framework:

MoSCoW (Must, Should, Could, Won't). The team huddled around the feature board as Maria grabbed a marker. *"Must Have: Log meals, walks, and meds...non-negotiable for launch,"* she began, tagging those core logging features as **Must**. *"Should Have: Reminder customization...helpful but not core,"* she added, marking the basic reminder feature as Should (important, but if it slipped to the next release, the app would still function). *"Could Have: maybe a profile picture upload,"* which they all agreed was nice-to-have eye candy, not essential. *"Won't Have (for now): Social login,"* Maria concluded, noting that letting users sign in via Facebook or Google could wait, as it added complexity with little impact on MVP goals.

Seeing these labels appear on the board brought immediate clarity. *"I can see the MVP backbone now,"* Jenna remarked, relieved that the core of the app was clearly distinguished from the extras. The MoSCoW exercise took only a few minutes but paid dividends in alignment. Everyone visually grasped which features were mission-critical. Maya reminded them that this was a living prioritization, *"It doesn't mean our decisions are final, but it gives us a direction."*

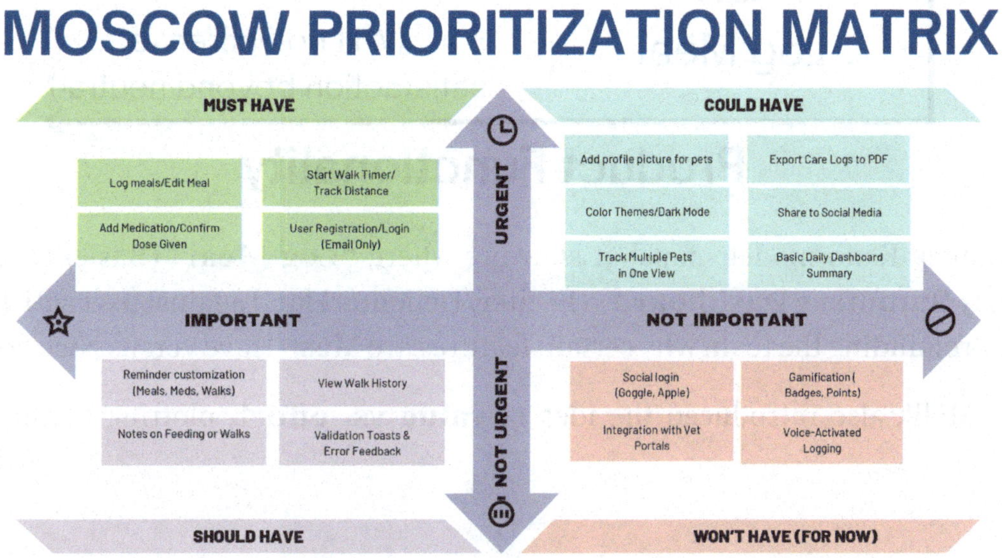

The team also discussed other prioritization lenses: Maya briefly explained the **Kano model** (to consider user delight). For example, a "Daily Summary Dashboard" might be a **Delighter** (users would love it, but they expect the app to function without it), whereas the ability to log a meal is a **Basic Need** (if it's missing, users are upset). Considering these helped reaffirm why some features were Musts (they met basic expectations) and others could wait (they'd delight but not deliver core value immediately).

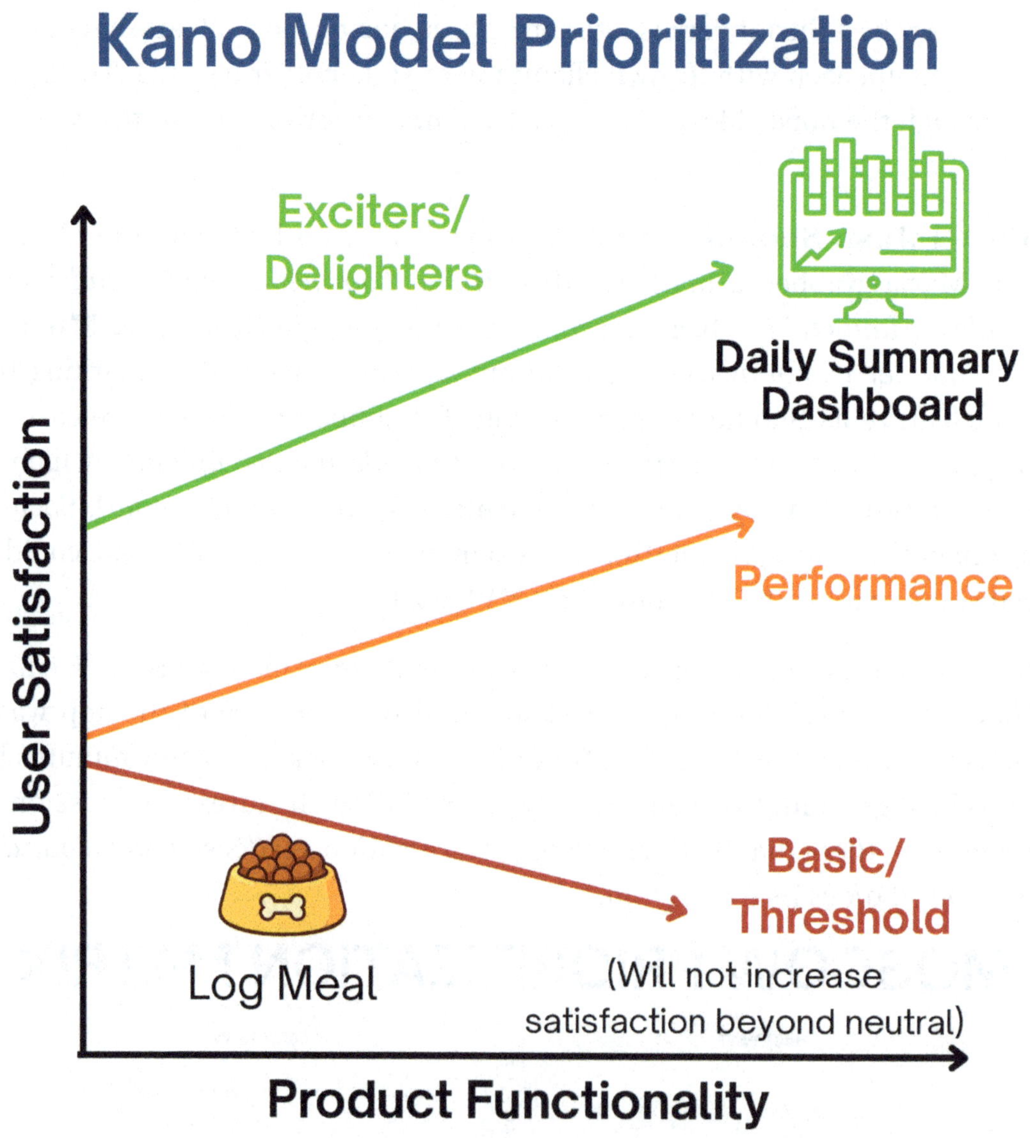

Here's a quick Kano-model snapshot showing where **"Log Meal"** (Basic/Threshold) and **"Daily Summary Dashboard"** (Exciter/Delighter) land against user satisfaction. Useful for reminding the team why certain features are Must-Have versus Nice-to-Have.

Alex and Malik also introduced the idea of **value vs. effort**, plotting features on a matrix to ensure the team tackled high-value, low-effort items first. In practice, most of the MVP features fortunately fell into a high-value category (feeding log was a non-negotiable value), and none seemed outrageously difficult technically (since they had spiked on any uncertain tech). After these discussions, each team member felt a lot more clarity about *"what comes first and why."* Maria noted she could now defend the backlog priorities to any VP or stakeholder with logic and user evidence. Alex was happy to focus on the most impactful code first. Jenna saw which UI to perfect now versus later. Malik knew where testing had to be rock solid versus areas that could be explored in beta.

The end result: Team Vertex distilled a chaotic list into a **focused short-list** for Sprint 1: things like the Registration/Login flow and Profile setup, the absolute Musts, would be done first.

- **Product Roadmap; Sequencing the Sprints:** By the end of requirements analysis, the team not only knew *what* to build for MVP, but also had a rough plan for *when* to build each part. To solidify this, they created a lightweight **Product Roadmap** laying out the next few Sprints in broad strokes. This roadmap was an outcome-driven schedule, not a fixed date plan, it answered *"what order will we tackle these epics/features in?"* Maya facilitated this roadmap discussion, ensuring it aligned with their earlier decisions.

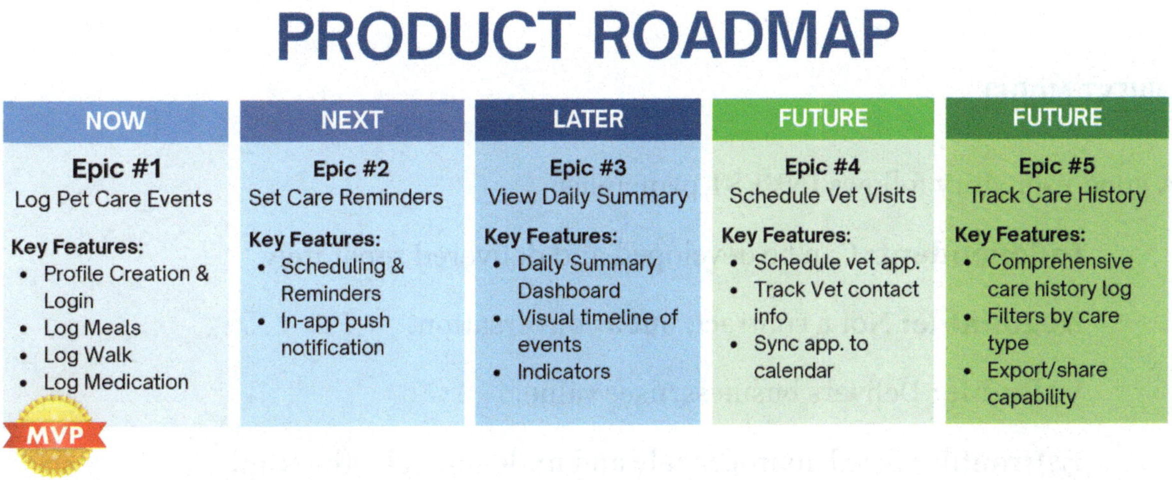

The team organized the roadmap by **timeframes** rather than exact dates, using buckets like "Now (current Sprint), Next, Later, Future." They filled in each timeframe with the outcomes, epics, and key features they expected to deliver:

Key Learning Points:

Before development begins, Agile teams rely on clear, concise, and testable user stories to describe the features they aim to deliver. This practice anchors communication between stakeholders, developers, testers, and designers.

USER STORIES

Purpose: Capture user needs from the user's perspective in a lightweight, goal-oriented way.

Format: As a [user], I want [action] so that [value].

Value Focus: They keep the focus on user value and encourage empathy and alignment.

ACCEPTANCE CRITERIA

Purpose: Define the scope and behavior of a story.

Format: Gherkin syntax, GIVEN [context], WHEN [action], THEN [outcome]

Alignment Benefits: Clarifies shared understanding and provides a basis for testing.

INVEST MODEL

A good user story follows INVEST principles:

Independent: Can be developed and delivered separately.

Negotiable: Not a contract, but a conversation.

Valuable: Delivers business/user value.

Estimable: Sized appropriately and understood by the team.

Small: Can be delivered within a Sprint.

Testable: Has clear acceptance criteria.

3 C'S MODEL

Card: A story written down (usually in a backlog system like Jira or Trello).

Conversation: Ongoing discussions to refine the story.

Confirmation: Agreement on the definition of done and acceptance criteria.

This roadmap encapsulated the team's plan on one page. It showed how the Feeding feature (part of "Core Activity Logging") was slated for immediate development ("Now"), whereas things like the Daily Summary (which includes feeding history visualization) would come a bit later. The roadmap provided a sanity check: it *aligned the team's short-term work with long-term goals.*

It also became a communication tool. Maria could show this to stakeholders and explain, for example, that **Feeding and other logging features are coming in the first release**, while the fancy dashboards are scheduled for a later iteration. Importantly, the roadmap was **flexible**. Maya emphasized that it was *"direction, not a deadline."* If priorities needed to shift due to user feedback or new insights, they would adjust the roadmap rather than treat it as a rigid contract. Nonetheless, having this sequence laid out quelled a lot of internal anxiety.

Team members now understood how their work in Sprint 0 (preparing requirements for Profile Creations, etc.) fed into Sprint 1 and beyond.

As Alex said, "If I know Sprint 1 will focus on Login & Profile Creations to prove our logging flows, I can architect the backend with that in mind, making room for future summary features without a rewrite."

Jenna agreed, noting she would prioritize designing the Login and Registration screens first since those would be used immediately, and save the dashboard design for later.

Malik too saw benefit: he could start writing tests for profile creation now and hold off on tests for future epics until they neared "Now" on the roadmap. In essence, the roadmap exercise shifted Team Vertex from a *reactive* mode (just refining a backlog in isolation) to a *proactive* mode of thinking ahead. It gave context to every user story they would write next, ensuring each tied back to a bigger picture outcome.

By the conclusion of Sprint 0's Requirements Phase, Team Vertex had transformed a vague feature idea, Profile Creation/Login/Register & **Feeding log**, into a structured plan. They understood who needed it (Emily and pet parents like her), what it should do (enable quick meal logging and tracking), how it would work (via defined workflows and UI steps), and why it was important (to ease a real user pain point). All of this was documented in their **product backlog** and design artifacts, ready to guide the next steps of development.

AGILE IN MOTION

Throughout this phase, the team actively practiced Agile principles and SDLC activities, making the process just as important as the outcome:

- **Stakeholder Workshops & Interviews:** The persona creation workshop led by Maria was grounded in real user input. The team used data from prospective users (pet owners and vets) to shape personas and uncover true needs. Later, they would also conduct Joint Application Design (JAD) sessions where stakeholders gave feedback, ensuring that requirements stayed aligned with stakeholder expectations. Engaging users and stakeholders early through interviews and feedback sessions ensured that the Feeding feature was defined with *empathy*. It addressed genuine pain points (like Emily's forgetfulness about feedings) rather than assumptions. This collaborative requirement gathering reflects Agile's emphasis on customer involvement.
- **Process Modeling:** Team Vertex employed **visual modeling techniques** (use case, workflow, and activity diagrams) to clarify requirements before coding. This aligns with the SDLC **Requirements Analysis** best practices, modeling processes to identify gaps and confirm understanding. By drawing the **"Log Feeding" workflow and activity diagrams**, the team achieved a shared mental model of how the feature should function, bridging any gaps between the business view (what the user needs) and the technical view (what to implement). These models acted as communication tools: rather than lengthy text specs, a simple diagram conveyed the flow to both developers and non-technical stakeholders. Agile encourages "just enough" documentation, and these diagrams were lightweight but extremely informative, evolving as needed (the team was prepared to update them if new edge cases emerged during development).

- **Incremental Story Mapping:** The user story map the team built is a core Agile practice to maintain a **user-centric backlog**. It allowed them to prioritize **flows over isolated features**, making sure the Feeding feature was considered in context. This activity embodies the principle of **iterative requirements refinement**: they first listed big activities, then refined them into stories, then pruned those stories to an MVP. Each step (personas → use cases → workflows → stories) narrowed the scope and increased detail in an iterative fashion, which is the essence of Agile requirements engineering.
- **MoSCoW Prioritization:** The use of MoSCoW categories (Must/Should/Could/Won't) is a straightforward yet powerful Agile prioritization method. Team Vertex used it during backlog refinement to fight the urge to do everything at once. By explicitly categorizing Feeding and other features as Must or Should, they created a **shared understanding of priority**. This prevented later conflicts and scope creep. It also ties to the idea of **timeboxing**; they knew if something was a "Could" or "Won't" for MVP, it would automatically be timeboxed out of early Sprints. The team also complemented MoSCoW with **Kano analysis** (thinking about user delight) and a bit of **cost-benefit thinking** (value vs. effort). This mixed approach ensured a well-rounded prioritization that considered user impact, business value, and development effort, all Agile alignment practices.
- **Continuous Team Collaboration:** All six team members were involved in defining requirements, each bringing their perspective. This cross-functional teamwork is a hallmark of Agile. For instance, when breaking down epics into features, *Jenna* ensured the user's perspective was represented ("do we need a profile photo upload for a better UX?"), *Alex* considered technical simplicity ("let's stick to email/password login initially"), *Malik* raised quality concerns ("we need to include validation stories for input errors"), and *Maya* facilitated and kept the process efficient. The result was a set of requirements that was balanced and agreed upon by all disciplines, reducing misunderstandings down the line. Agile in motion means the **conversations** around user stories are as important as the written stories themselves, and Team Vertex truly embraced that: every user story they wrote was discussed and clarified as a group, following the "three C's" of a good story (Card, Conversation, Confirmation).

Breaking Features into User Stories

To support iterative development, epics and features must be broken into testable, value-driven user stories. Below is a worked example using core features from the PetTracker MVP.

Feature: Log Meal

User Story: As a pet owner, I want to record what my pet ate so that I can track their diet.

Acceptance Criteria:

GIVEN I'm on the pet profile, WHEN I select "Log Meal" and enter food details, THEN it should be saved to the pet's record.

GIVEN I forget to enter food details, WHEN I tap save, THEN I should see an error prompting me to complete the required field.

Feature: Log Walk

User Story: As a pet owner, I want to log my pet's walks so that I can monitor their daily activity.

Acceptance Criteria:

GIVEN I open the Walk Log form, WHEN I select the date, duration, and distance, THEN the walk should be stored under the correct pet.

GIVEN I submit a walk entry, WHEN data is saved, THEN I should receive a confirmation message.

Feature: Log Medication

User Story: As a pet owner, I want to track my pet's medication doses so that I don't miss any critical treatments.

Acceptance Criteria:

GIVEN I enter the medication name and time, WHEN I tap save, THEN it should appear in the pet's history.

GIVEN I set a reminder, WHEN the scheduled time arrives, THEN I should receive a notification.

These stories guide design, development, and testing. They also reinforce vertical slicing, delivering a thin, functional slice of value end-to-end.

Theme vs. Epic vs. Feature vs. User Story

Level	Definition	Scope	Example (PetTracker App)
Theme	A high-level strategic objective that ties related Epics together under a common goal.	Broad, long-term direction	Give pet parents a simple way to log, manage, and visualize daily care routines while reducing missed tasks.
Epic	A large body of work that delivers a distinct capability and can be broken down into Features and Stories.	Major capability; spans multiple sprints	Log Pet Care Events: covers meals, walks, and medications.
Feature	A specific deliverable within an Epic that provides end-user value.	Medium-sized, can often be delivered in 1–2 sprints	Logging Meals: select pet, choose meal type, add time, and notes.
User Story	A small, testable requirement written from the end user's perspective.	Granular; completed within a sprint	As a pet parent, I want to log my pet's meals so I can keep track of feeding times.

This chart illustrates the hierarchy of work in Agile, moving from the broadest vision down to the most actionable tasks.

- At the **Theme** level, the team defines the *strategic objective*, the "why" that guides the product direction.
- **Epics** represent *major capabilities* that deliver user value and can be broken down into smaller parts.
- **Features** describe *specific functionalities* within an Epic that can be delivered incrementally in one or two sprints.
- **User Stories** are the *granular requirements* written from the end-user's perspective, providing clear, testable work items for the team.

In the PetTracker example, the Theme focuses on helping pet parents manage daily care routines. Within that Theme, the Epic **"Log Pet Care Events"** captures the high-level capability of recording meals, walks, and medications. A feature like **"Logging Meals"** narrows the scope to a single function, which is then expressed in actionable **User Stories**, such as logging a pet's meal with details like type, time, and notes.

In summary, this Requirements Phase exemplified Agile methodology: customer-focused, iterative, and collaborative. The Feeding feature was not defined in a vacuum or via a giant spec document handed down; it was **co-created in real time** by the team using Agile tools and mindset.

Estimation & Story Pointing

In Sprint 0, once the initial backlog is sketched, the team establishes a shared sense of effort, complexity, and risk for each item. Estimation here isn't about exact time; it's about building a common understanding so Sprint Planning is realistic.

- **Backlog Refinement & Grooming:** Even before formal development Sprints, the team engaged in backlog refinement sessions. They **added details to user stories**, identified acceptance criteria, and even performed initial **estimations** using Planning Poker for a few key stories. For example, they pointed out the Feeding Log story at 5 points after discussing complexity. These activities are part of SDLC Requirements Analysis (detailing out requirements) but done in an Agile way: collaboratively, just-in-time, and with the whole team's input. The presence of QA (Malik) and Dev (Alex) in these discussions meant that test scenarios and technical constraints were considered early, which improves quality and feasibility of requirements.

WHY STORY POINTS (NOT HOURS)?

STORY POINTS ARE A RELATIVE SIZE THAT BLENDS COMPLEXITY, EFFORT, AND UNCERTAINTY. A "5" IS BIGGER THAN A "3," REGARDLESS OF WHO BUILDS IT. THIS KEEPS THE FOCUS ON COMPARATIVE DIFFICULTY, NOT CLOCK TIME.

Common Scales

Fibonacci: 1, 2, 3, 5, 8, 13... (uncertainty grows with size)

T-shirt sizes: XS, S, M, L, XL (useful for high-level sizing)

Estimation Techniques:

- **Planning Poker:** Everyone selects a size privately, reveals together, discusses gaps, and converges.
- **Affinity Estimation:** Quickly sort many stories from smallest → largest by comparison.
- **Bucket System:** Drop stories into predefined "size buckets" to size a big backlog fast.

Spikes for Unknowns

When a story has significant unknowns (e.g., GPS background tracking, push notifications), create a time-boxed spike to research or prototype first. Use the spike's findings to size the actual story more accurately.

Best Practices

- **Estimate as a team** (Dev, QA, UX, PO input = better clarity).
- **Break down large stories** until they fit comfortably in a Sprint.
- **Avoid anchoring** (simultaneous reveal helps).
- **Revisit estimates** when scope or understanding changes.

This light-touch sizing in Sprint 0 gives Team Vertex a shared yardstick for later Sprints (e.g., Feeding Log, Walk Tracking, Medication Logging) and reduces surprises when planning.

TEAM REFLECTION

The Sprint 0 requirements work was intense, but it left Team Vertex stronger and more aligned. They reflected on a few key insights gained:

- **User-Centric Mindset:** Maria (PO) observed that developing personas and story maps made her see the product "through the user's eyes, not just the backlog's." She realized how crucial it is to ground every decision in user value. Whenever the team wavered on a feature, referring back to Emily's persona or her story map of daily routine clarified what to do. This user-first perspective became a guiding principle moving forward.
- **Shared Understanding:** Jenna (Designer) felt that visual techniques

(like the diagrams and mapping) got everyone on the same page. *"It helped me design the real journey, not just isolated screens,"* she noted. Alex (Developer) added that now he knew "what flow to code and which endpoints to build." The exercise of mapping and discussing flows gave him context that pure requirements text might not have. Malik (QA) appreciated understanding the flows and edge cases early: *"I can build test cases around use paths and fields now,"* he said, meaning fewer surprises during testing.

- **Confidence in Priorities:** After applying prioritization frameworks, the team felt more confident and less anxious about the order of execution. *"I could defend my choices with logic and intuition,"* Maria remarked, relieved that she could justify to higher-ups why feeding logs and basic reminders come first. Alex added, "It helped us focus on what would deliver value quickly without overcomplicating the build," confirming that the team could create impact early without unnecessary effort. The once overwhelming list of possible features now had a sensible structure and sequence.
- **Adaptability and Continuous Learning:** Team Vertex also realized that this phase was just the beginning of learning. By doing spikes and prototyping (Jenna's Figma mocks, Alex and Malik's code spikes on authentication and timers), they pre-emptively tackled some uncertainties. The team acknowledged that they must stay adaptable; if new information comes up in Sprint 1 (say, user testing reveals a flaw in the profile creation flow), they are prepared to revisit requirements. In reflection, Maya (SM) reminded everyone that **priorities and understanding will evolve**, and that's okay. The strong foundation they built in Sprint 0 will make those inevitable changes easier to handle because the team is now cohesive and **speaking the same language** when it comes to the product.

Overall, the requirements phase not only produced a backlog and diagrams but also fostered a culture of collaboration and clarity. The team is now **in sync** about what they're building and why, a unity that will be invaluable in the Sprints to come.

Exercises & Activities

Quick Check

Take a moment to reflect on the work done in Sprint 0's Requirements Phase. Consider these questions to check your understanding and stir your critical thinking:

- **Backlog Building:** How did creating visual maps (like workflows and story maps) change the way you think about building a backlog? Did it help you spot missing pieces or priorities more clearly?
- **Early Epics:** How might defining epics early in the process help with estimating effort, planning release phases, and keeping stakeholders aligned on scope?
- **MVP Mindset:** In your own words, how does thinking in MVP terms help a team avoid feature creep and stay focused on what matters first?
- **Emotional vs. Logical Priority:** During prioritization, were you surprised by any feature's priority? Did any feature feel more important emotionally than it turned out logically? (For example, a flashy idea you loved but realized wasn't crucial for MVP.) How does a structured approach help separate true user value from "coolness"?
- **Communication Challenge:** If a stakeholder suggested a new feature during this phase (say, a pet social feed), what guiding questions would you now ask to decide if it's in scope or a later consideration? Reflect on how you would explain the MVP vs MMP decision to them to manage expectations.

Write down your thoughts or discuss with a peer. These reflections ensure you're internalizing not just *what* Team Vertex accomplished, but *why* those steps were important.

Apply This

Put yourself in Team Vertex's shoes with these scenario-based exercises. Use the following activities to apply the concepts from this chapter to a project of your own (or a fictitious one):

- **Part 2: Break Each Activity into User Stories**. For each of the 3 activities, list

EXERCISE 1: CREATE A USE CASE DIAGRAM

Create a UML Use Case Diagram that illustrates the interaction between Pet Parents and Support Team Members and the PetTracker system.

Instructions

Identify the Actors

- Determine who interacts directly with the app and who supports its operation behind the scenes.

(Hint: Think "Who initiates actions?" not "Who's affected by them.")

List the Use Cases

Define what actions each actor can perform within the system. Each use case should represent a meaningful goal from that actor's perspective.

Draw the Diagram

Create a rectangle labeled PetTracker to represent the system boundary.
Place the actors outside the box (using stick figures).
Inside the box, add ovals for each use case and connect them with solid lines to the actor(s) that perform them.

Label Everything Clearly

- Use the exact use case names provided (don't rename them).
- Optional: Add short notes or include relationships like «include» or «extend» if they fit logically.

Reflect on the MVP

- After completing your diagram, identify which use case you would prioritize for Sprint 2 and why. (Clue: Think "daily user value.")

Hint from Team Vertex

They realized that mapping out these use cases early prevented scope confusion later during backlog refinement. It also helped developers and testers clarify what "done" means for each user interaction.

Deliverable

- A single Use Case Diagram (hand-drawn, digital, or tool-based) that includes:
- Two actors: Pet Parent and Support Team Member
- Six use cases matching the PetTracker scope
- A short note naming the most critical use case and your reasonin

Exercise 2: Build a Workflow & Activity Diagram

Scenario: You're in Sprint 0 with Team Vertex. The feature **"Log Feeding"** is next in the pipeline. Your job is to visually map the user flow using both a Workflow Diagram and an Activity Diagram to clarify the interactions **before** writing stories.

- **Part 1: Build a Workflow Diagram (High-Level Steps):** Using boxes and arrows, sketch a horizontal workflow diagram for how a pet parent would log a feeding in the app. List **6–8 core steps** from start to finish (no need to detail conditional logic yet). *(Hint: Think about the major screens or actions, from opening the app to seeing the confirmation.)* Write out the steps in order. For example, you might start with "User opens app" -> "Navigates to Log Feeding screen" -> ... and so on.

- **Part 2: Expand into an Activity Diagram (With Decisions)**: Now take your basic flow and add decision diamonds for at least **two validations** or alternate paths. Identify points where the user's input could branch the flow. For each decision, specify what happens on the **Yes/No** or **Success/Failure** paths. Also, think of one **parallel action** if applicable (e.g., a background process or spinner). Questions to guide you: What should the app do if required info is missing (no pet selected, no meal time)? What happens after the user hits "Submit" (both on success and on error)? Sketch or describe these branches in your activity diagram.
- **Part 3: Derive Stories & Spikes**: From your diagram, list out at least **2 user stories** and **1 spike** that you would create for the backlog. Also, come up with **1 important validation rule** that would likely become an acceptance criterion. For example: a story might be "As a user, I want to select a meal type from predefined options," and a spike might be "Research local notification scheduling for feeding reminders." A validation rule could be "portion amount must be a positive number." Write these down.
- **Part 4: Link Back to Agile Artifacts**: Describe how you would reference or use these diagrams once you start implementation. Where would you save or display them for the team's benefit? How would having this diagram change conversations during refinement or stakeholder meetings? (E.g., "I'd attach the workflow diagram to the user story in Jira so any new team member or stakeholder can immediately visualize the user journey.")

By completing this exercise, you practice turning a feature concept into visual documentation and backlog items, just like Team Vertex did.

EXERCISE 3: BUILD YOUR OWN USER STORY MAP

Scenario: You're joining Team Vertex as they plan the MVP for PetTracker. Thevision is clear: Help pet parents log and manage daily care routines. Now it's yourturn to lead the User Story Mapping session.

- **Part 1: Define the User & Activities.** First, identify who your primary user is and what high-level activities they need to accomplish. In PetTracker's case, the User Persona is Emily (a busy pet parent caring for Luna). List 3 high-level user activities Emily performs in her daily pet care routine (these will become your horizontal swimlanes). For example, in the story above, the activities were Feed, Walk, and Medicate. Come up with three major categories of tasks for your persona.

2–3 user stories that describe specific tasks the user would do (and why). Focus on the **needs** of the user. For each activity, mark at least two stories as "Must" (critical) and one as "Could" (nice-to-have) for the MVP. You can format it like:

Activity 1:

Must: *As a* ________, *I want to* ______________________________ *so that* _______________.

Must: *As a* ________, *I want to* ______________________________ *so that* _______________.

Could: *As a* ________, *I want to* ______________________________ *so that* _______________.

Do this for Activity 2 and Activity 3 as well.

- **Part 3: Slice Your MVP.** Imagine drawing a line under the last "Must" story for each activity. Now list out all the "Must" stories from across your map; these form your MVP slice. Write the list of MVP stories (the ones you would include in the first release). How many are there? Does this set cover the core needs of the user persona? (It should!).
- **Reflection:** How did mapping the user's journey influence your perspective on what's important? Did you notice any story that didn't really fit the main activities (and thus could be dropped or moved down in priority)? Write a few notes on how this exercise changed your approach to thinking about features.

EXERCISE 4: GROUPING STORIES INTO FEATURES

After creating a bunch of user stories, it's time to group them into epics like Team Vertex did. **Scenario:** You're reviewing a portion of Team Vertex's user story map. Below are 10 user stories (mixed from various activities). Your task is to **group them into logical epics** and give each epic a clear name, similar to how Team Vertex identified *Meal Logging*, *Walk Management*, etc.

Here are the user stories:

1. *As a user, I want to log a meal with food name and portion size, so that I can accurately track my pet's nutrition.*
2. *As a user, I want to receive a reminder when it's time to walk my pet, so*

that I never miss scheduled exercise and keep my pet healthy.

3. *As a user, I want to set my pet's medication schedule so that doses are given consistently and on time.*
4. *As a user, I want to update my pet's profile photo so that the profile always reflects my pet's current appearance.*
5. *As a user, I want to view my pet's feeding history by day, so that I can monitor eating patterns and spot any irregularities.*
6. *As a user, I want to start and stop a walk with a timer, so that I can record the exact duration of each walk.*
7. *As a user, I want to get an alert when it's time to give medication, so that I don't miss a dose and keep treatment on track.*
8. *As a user, I want to edit or delete a logged meal, so that my feeding records remain accurate if I make a mistake.*
9. *As a user, I want to view past walks on a map, so that I can review routes and vary them for my pet's enrichment.*
10. *As a user, I want to enter my pet's name, age, and breed during profile setup, so that the app personalizes care recommendations for my pet.*

Now, group these into 3–4 Features. For example, stories 1, 5, and 8 might belong to an epic called "Meal Logging & Tracking". List your Epic names and which story numbers go under each. You might end up with epics like:

- 🐾 **User & Pet Profile Setup**: (stories _____)
- 🍽️ **Meal Logging & Tracking**: (stories _____)
- 🚶 **Walk Activity Management**: (stories _____)
- 💊 **Medication Scheduling & Alerts**: (stories _____)

(The emojis are just for fun, to symbolize each epic's theme, as seen in Team Vertex's board.)

Optional Reflection: How might defining epics early help the team during estimation and release planning? Think about how having those "buckets" could make it easier to communicate the plan to stakeholders or to divvy up work among team members.

EXERCISE 5: MVP OR MMP? DECIDE WHAT SHIPS FIRST

Team Vertex had to differentiate what features were MVP-critical and which could wait for the MMP. Try this with a sample list: Scenario: You're part of Team Vertex planning the PetTracker app's first release. You have a list of potential features. Decide which ones are **MVP (Must-Have for launch)** and which are **MMP (can come later)**:

Part 1: Sort These Features

Feature	MVP or MMP?
Log a meal with timestamp	
Select a walk type (e.g., park, neighborhood)	
View daily summary dashboard	
Push notifications for medication reminders	
Add pet profile photo	
Share care log with roommate	
Export care history to PDF	

Go through each feature and mark it as "MVP" or "MMP." (There's no single correct answer, but think in terms of core functionality vs. enhancements.) For instance, "Log a meal" is almost certainly MVP, since that's core logging. "Profile photo" is likely MMP, since it's not functionally critical.

Next, **justify two of your choices**: pick one feature you labeled MVP and explain *why* it absolutely needs to ship in the first version. Then pick one feature you labeled MMP and explain *why* it's okay to delay. This mirrors Maria and Alex's thought process when they argued about what the MVP should contain.

Finally, consider: **How does having an MVP mindset help avoid feature creep?** Jot down a sentence or two about how explicitly deciding MVP vs MMP features keeps a team focused (for example, it gives a clear "not now" list for anything not MVP).

EXERCISE 6: FROM VISION TO BACKLOG. BUILD YOUR FIRST

AGILE BACKLOG

Let's simulate the initial backlog creation for a vision. **Scenario:** You've just joined Team Vertex in Sprint Zero. Maria (PO) has presented the PetTracker MVP vision: *"Empower pet parents to log, manage, and visualize their pets' daily care routines, while reducing missed tasks like medication or vet visits."* Your task is to help convert that vision into a structured, testable **Product Backlog**, starting with Epics, then Features, and finally mapping them to Sprints.

- **Part 1: Define 2 Epics**. Based on the vision above, propose two high-value epics that would be essential to fulfilling that vision. (For example, one might be related to logging care events, another to reminders or summaries. Use the vision keywords as hints: "log/manage routines" might be one epic, "visualize routines" might be another.) Write Epic 1 and Epic 2 in one sentence each.
- **Part 2: Decompose Each Epic into 3–4 Features**. For each of the epics you named, list three or four features that would fall under it. Remember, a feature is something that can be built in a Sprint and delivers on part of the epic. For instance, if Epic is "Care Event Logging," features could be "Meal Logging," "Walk Logging," "Medication Logging," "Editing Logs," etc. Do this for both epics.
- **Part 3: Assign Features to Sprints**. Now imagine you have to schedule those features into the first 2 or 3 Sprints. Consider logical order and any dependencies. Create a mini plan for **Sprint 1, Sprint 2, and Sprint 3**, listing which feature(s) would likely go into each. (Sprint 1 might take foundational features, Sprint 2 builds on that, etc.)
- **Part 4: Write a Sample Acceptance Criteria**. Take one feature from your list and write 2–3 acceptance criteria for it (using a format like Given/When/Then or simply bullet points of conditions). For example, if the feature is "Log a Meal," an acceptance criterion might be "Given the user has input a food name and portion, when they tap Save, then the meal entry is saved to the log with a timestamp." Focus on capturing the core *behavior or rule* that would let a tester know "this feature works as intended."

By completing this end-to-end exercise, you practice the *entire arc* of requirement analysis: from a broad vision to specific testable pieces. This mirrors what Team Vertex achieved by the end of Sprint 0, a backlog blueprint ready for Sprint 1.

EXERCISE 7: PRIORITIZE THE CHAOS...WHAT COMES FIRST?

Prioritization is tough when everything seems important. Here's a scenario to test your prioritization skills: **Scenario:** You're stepping into Maria's shoes as Product Owner. Team Vertex just finished a session and now has 8 user stories on the table. Only a few can fit in the next Sprint. Use prioritization frameworks to bring clarity.

The 8 user stories are:

- Log a pet's meal
- Add profile picture
- Set a reminder for medication
- Display visual pet timeline (activity history graph)
- Allow editing walk entries
- Social media sharing of pet activities
- Customize notification sounds in the app
- Add vet visit notes to pet profile

- **Part 1: Apply MoSCoW**. Go through each story and assign a priority: Must Have, Should Have, Could Have, or Won't Have (for the MVP). You can create a simple table or list, for example: "Log a pet's meal: Must Have; Add profile picture: Could Have; ..." etc. Remember, "Must" means essential for the MVP or next Sprint, "Won't" means definitely not now (maybe later). Be realistic and try to have only a few "Musts" (since not everything can be a top priority).
- **Part 2: Apply Kano Analysis**. Now take the same list and label each story as a Basic Need (users expect it), a Delighter (would wow users but they wouldn't miss it initially), or Neutral (neither expected nor particularly exciting). For instance, "Log a pet's meal" is a Basic Need (core functionality), while "Social media sharing" might be a Delighter or even Neutral (some users might love it, others might not care). This perspective helps you double-check your MoSCoW choices against user satisfaction. Note down your Kano category for each story.
- **Part 3: Reflect**. After doing the above, answer: Which story was hardest to prioritize and why? Did the Kano perspective make you rethink any of your MoSCoW assignments (e.g., realizing something you put as "Should" is actually a Basic Need and should perhaps be a "Must")? How does using a structured approach like this help when team members or stakeholders each push for their favorites? Write a brief reflection.

This exercise gives you a taste of how Team Vertex used MoSCoW and Kano to debate and decide on their backlog priorities. It's a skill that will come in handy whenever you have more ideas than you can build (which is almost always!).

EXERCISE 8: BUILD YOUR OUTCOME-BASED ROADMAP

Finally, practice creating a mini product roadmap Agile-style: **Scenario:** You're joining Team Vertex early in the PetTracker project. The backlog is full of features, but the team lacks a big-picture timeline. Help sketch a **lightweight Agile roadmap** to bring sequence and purpose to the plan.

- **Part 1: Define Themes (Outcomes).** Think of 2–3 high-level *themes or outcomes* PetTracker wants to achieve. These are the "Why" buckets. For example, outcomes could be *Increase user engagement*, *Improve care tracking accuracy*, or *Reduce missed care events*. List your chosen themes.
- **Part 2: Group Features into Epics under Themes.** Take some sample features (you can reuse from earlier exercises or invent a few, like "Log a meal," "Schedule feeding reminders," "View daily summary," "Track care history," "Push med notification," "Export care data"). Now group these into 2 epics that align with the themes. For instance, under a theme "Improve care tracking accuracy," you might group "Log a meal" and "Schedule feeding reminders" into an epic called *Core Care Logging*, and under the "Increase engagement," group "View daily summary" and "Share care data" into an epic called *Care Insights Sharing*. Create **Epic 1** and **Epic 2**, and list a few features under each that belong there.
- **Part 3: Assign Timeframes.** Decide an order for your epics. Which epic should be tackled Now, Next, Later, and so on? Draw up a mini roadmap table or list with columns: Timeframe (Now/Next/Later), Outcome/Theme, Epic, Key Features. Fill it in with your two epics and any others you think Team Vertex would have, arranged in a sensible sequence. For example: **Now:** Outcome = Core Care Logging, Epic = Log Pet Care Events (Features: Log Meals, Log Walks, Log Meds) **Next:** Outcome = Proactive Care, Epic = Set Reminders (Features: Feeding reminders, Walk reminders, Med reminders) **Later:** Outcome = Insights & Sharing, Epic = Daily Summary Dashboard (Features: summary view, timeline, export data)
- **Part 4: Reflect.** How would having this roadmap have changed the conversations

in Team Vertex's requirement workshops? Who on the team benefits most from seeing the roadmap early (PO, devs, QA, etc.)? Also, consider how you would keep such a roadmap useful yet flexible. What would you do if a new high-priority outcome emerges?

By completing this, you emulate the roadmap exercise Team Vertex undertook. A good roadmap keeps everyone oriented towards **common goals** and helps in making tough choices when new requests pop up.

EXERCISE 9: WRITING USER STORIES THAT ACTUALLY WORK

Learn to identify what makes a user story poorly written and practice rewriting it using the INVEST principles and the As a [role], I want [goal], so that [value] format.

INSTRUCTIONS:

Analyze the Story

- Read: "As a user, I want to log data."
- Ask yourself:

  Who exactly is this user?

  What kind of data are they logging?

- Why does it matter?

Identify the Gaps

- Note everything unclear, missing, or too generic about the story.
- Think from both a user and tester perspective—could this story be developed or tested as-is?

Rewrite the Story

Use the correct format:

As a [specific user role], I want [clear action or goal], so that [defined value or benefit].

- Make sure your version reflects a real user scenario relevant to the PetTracker app.

Check for Clarity

- Could someone new to the project understand the story at a glance?
- Does it show purpose and user value?

Reflection Bonus (Optional)

Imagine a stakeholder submits another vague story like: "As an admin, I want to store system logs."

How would you coach them to make it more user-focused and business-relevant?

Write your short response below (1–2 sentences).

Exercise 10: Define It Right – Acceptance Criteria in Action

Scenario: The PetTracker team has just completed their first user story for Sprint Zero: "As a pet parent, I want to log my pet's medication so that I can keep track of doses and avoid missed treatments." Everyone nods—it's a solid story. But before this story can move to "Ready for Development," the Scrum Master reminds the team: "A user story without clear acceptance criteria is just a wish list." Now it's your turn to help Team Vertex transform this story into something developers and testers can actually work with.

Objective

Define clear, testable acceptance criteria that describe exactly how the system should behave when the pet parent logs a medication.

Instructions

1. Start with Two Plain-English Criteria

Write at least two acceptance criteria in bullet-point form that describe what must happen for the story to be considered done.

(Tip: Think "what does success look like?")

2. Translate Them into Gherkin Format (Given/When/Then)

Convert your acceptance criteria into the BDD-style syntax that testers can automate later.

- Given sets the context
- When describes the action
- Then defines the expected outcome

3. Stretch the Thinking

Add three more conditions covering different angles:

- Functional: What must the feature do to meet business rules?
- Edge Case: What could go wrong that the system must handle gracefully?
- Non-Functional: How fast, secure, or reliable should the system perform?

4. Double-Check for "Ready"

When you're done, ask yourself:

- Could a developer build this confidently?
- Could QA test it without asking for clarification?
- Would the Product Owner know when to accept or reject it?

Reflection Bonus (Optional)

Imagine your Product Owner writes: "System should save data quickly."

How would you reframe that into a measurable, testable non-functional requirement?

Write your one-sentence rewrite below.

Role Replay

Communication is key in Agile. Practice a **role-play scenario** to solidify how Team Vertex might handle requirement discussions:

One common challenge is when stakeholders or team members propose features or requirements that are vague or overly technical. Imagine during a backlog refinement session, a stakeholder says, "We need to implement X algorithm for meal logging because I read it's the latest tech," without explaining the user value. As a team member, how do you respond?

Role Replay Scenario: *You* are the Product Owner (Maria) in a discussion with a stakeholder (could be a tech-savvy executive or a client). The stakeholder has suggested a feature request in technical terms (or something that doesn't tie clearly to user needs). Your goal is to guide the conversation to clarify the **user story** behind it or to gently push back if it doesn't align with the MVP goals.

Write a short dialogue between you (PO) and the stakeholder. For example:

- **Stakeholder:** "I think we should incorporate a blockchain ledger for the feeding logs."
- **You (PO):** "Interesting. May I ask, what user problem would that solve for pet owners?"
- **Stakeholder:** "Well, it's more secure."
- **You:** "Security is important. Our current approach is already secure for our needs. Do you feel users are asking for verifiable feeding log data, or is there a specific risk you're concerned about?"
- **Stakeholder:** *[perhaps realizes it's not a must-have]* "I just thought it'd be cool. But if it's not a user ask, then maybe it's not urgent."
- **You:** "We can note it down for the future. Right now, our priority is making sure Luna's feeding gets logged reliably. That's what Emily needs most."

Your dialogue can be serious or light but focus on **asking the right questions** and **keeping it user-centric**. The aim is to practice how the team maintains scope and clarity when confronted with potentially derailing ideas. This is the kind of proactive communication Team Vertex used to keep everyone aligned.

(Feel free to swap roles. Maybe it's Developer Alex raising a concern and PO Maria answering, etc. The key is demonstrating collaborative problem-solving in conversation.)

Your Toolkit

Team Vertex produced a number of useful artifacts during this phase. Here we provide simplified versions of those artifacts and checklists, which you can use as a toolkit for your own projects:

- **Persona Snapshot:** *(Pet Parent Persona: Emily)*
- **Name:** Emily Johnson, 29, Marketing Specialist (Tech-savvy)
- **Pet:** Luna, 3-year-old Labrador
- **Goals:** Log Luna's meals, walks, and meds in <2 minutes; get reminders for next care task; share daily summary with her roommate.

- **Pain Points:** Forgets if she fed Luna on busy days; uses sticky notes and loses track; worries if logs are accurate when she's away.
  Use your persona to keep requirements grounded. Refer back to "What would Emily want?" when debating features.

Use Case Diagram Key Elements:

- **Actors:** Who interacts with the system (e.g., Pet Owner, Vet)
- **Use Cases:** Actions or goals (represented by ovals, e.g., Log Feeding, Set Reminder)
- **Associations:** Lines connecting actors to use cases they participate in.
  This helps ensure you've thought of all major capabilities. If an expected use case is missing an actor or vice versa, revisit your requirements.
- **"Log Feeding" Workflow Outline:** (for reference or to compare with your own)
- **Open App**
- **Go to "Log Feeding"** (from home screen)
- **Input Meal Details** (pet, meal type, time, notes)
- **Submit** (save the log)
- **Confirmation** (success message)
- **Update UI** (last meal time shown)
  (Keep workflows high-level. They're great for explaining the feature at one glance.)

Activity Diagram Snippet (Fill Feeding Form):

- *Decision examples:*
  – If no pet selected → show error and ask to select
  – If time is in future → error "cannot log future time"
  – If notes > 200 chars → warn "too long" (trim or retry)
- *Parallel example:*
  – While saving entry to server, show loading spinner
- *Outcome:*
  – If save succeeds (200 OK) → show "Success" toast and update log
  – If save fails (error code) → show "Error, try again" and keep data on screen
  (Use this to double-check you covered unhappy paths in your user stories and tests.)
- **INVEST Checklist for User Stories:** Every user story should be: **Independent, Negotiable, Valuable, Estimable, Small, Testable.**

For example, *"As a pet parent, I want to log a meal so I can track feeding consistency."* Is it **Valuable** (yes, to the user's goal), **Testable** (yes, we can verify a meal is logged and appears in history), etc? If a story doesn't meet these, consider splitting or rewriting it.

- **Acceptance Criteria Example (Given/When/Then):**
  Story: As a pet parent, I want to save a feeding entry.
  Acceptance Criteria:
  – **Given** I have entered a food name and portion, **When** I tap "Save," **Then** the entry is added to the Feeding Log and I see a confirmation.
  – **Given** I left any required field blank, **When** I tap "Save", **Then** the system prompts me to fill the missing information (no entry is saved).
  – **Given** I'm offline, **When** I attempt to save, **Then** the app queues the entry and marks it as "Pending sync".
  (Clear AC like these ensure the team knows when a story is truly done.)

MoSCoW Prioritization Reminder:

- **Must Have:** Without it, the product isn't viable (e.g., "Logging a Feeding")
- **Should Have:** Important, but the product can function without it short term (e.g., "Edit feeding entry")
- **Could Have:** Nice to add if time permits (e.g., "Pet profile picture")

Won't Have (for now): Agreed to leave out this round (e.g., Social sharing) *Use this when your backlog feels too large. It's a quick sanity filter.*

Team Dialogue Tip: Before moving on from requirements, do a quick role-play (like the Role Replay exercise above) with your team: have someone play devil's advocate or stakeholder and see if you can all clearly justify the MVP cut and the requirements you've defined. If the team can collectively defend why the Feeding feature is defined the way it is and why it's scheduled when it is, you're in good shape!

Keep these tools in your toolkit as you move forward. Requirements can be complex, but breaking them down with visual aids, checklists, and open conversations makes them manageable and (dare we say?) even fun to tackle.

CHAPTER SUMMARY

In **Sprint 0: Requirements Phase**, Team Vertex accomplished an impressive feat: they transformed a broad idea (*"Let's allow users to track their pet care"*) into a well-defined set of requirements focused on a key feature: **Feeding**. By the end of this phase, the team had:

- **Identified their users and needs:** Through personas like Emily, they kept the project rooted in real-world problems (like forgetting pet meals). This ensured the Feeding feature and others directly address user pain points, not just assumptions.
- **Clarified workflows and edge cases:** Using use case, workflow, and activity diagrams, they visualized how a pet owner would log a feeding, step by step. This helped catch edge cases (e.g., invalid input, offline attempts) early and prompted the team to devise solutions (like error messages and validation rules). The Feeding workflow became a blueprint that all team members could reference, creating a shared mental model of how the feature should function.
- **Converted flows into backlog items:** Every part of the Feeding process was translated into backlog entries: user stories for each action/goal and spikes for any uncertainties. Nothing was left ambiguous; if it was in the diagram, it was accounted for in the plan. This resulted in a **fully groomed backlog** of user stories with clear acceptance criteria (for example, *"Given I enter valid food and portion, when I tap Save, then the log is saved and a success message appears"*).
- **Established priorities and scope:** The team distinguished **Must-haves vs. Nice-to-haves** for the MVP. Feeding (meal logging) was cemented as a Must-have epic in the MVP, whereas things like social sharing were explicitly marked out of scope for now. By doing MoSCoW prioritization and considering user delight factors, they prevented scope creep and set expectations for what the first release will include. This clarity means when Sprint 1 starts, everyone knows what to focus on first (no debates about "maybe we should do this extra thing now." Those decisions have been made).
- **Created a roadmap for delivery:** Finally, Team Vertex laid out a multi-Sprint roadmap linking back to user outcomes. They decided on the sequence: first build core logging (including Feeding) in the upcoming Sprints, then add reminders, then add summary views, etc. This **roadmap** gives context to each Sprint; they're not just doing tasks in isolation, they're building towards a larger

goal. It also sets them up to communicate with stakeholders about **when** to expect key features without committing to rigid dates.

- **Fostered team alignment:** Perhaps most importantly, this phase got the whole team aligned and confident.

They ended Sprint 0 Phase 2 with **a clear vision and plan**:

> **"BY THE END OF REQUIREMENTS ANALYSIS, WE HAD A FULLY GROOMED BACKLOG, ROLE-BASED USER STORIES, WORKING MOCKUPS, AND A SPRINT-READY ROADMAP".**

Everyone from product to QA understands the what and why of the Feeding feature and beyond, which means fewer misunderstandings later. The team also grew in their understanding of Agile practices, as many had their first real experience with story mapping, MoSCoW, and collaborative backlog creation, and they saw the value of these firsthand.

In short, Sprint 0's Requirements Phase set a strong foundation. The **Feeding feature** went from concept to concrete: its functionality, boundaries, and importance are now well-documented and agreed upon. Team Vertex is now ready to transition into design and development with confidence, having de-risked the project through careful planning and analysis.

SDLC PHASE 2: REQUIREMENTS ANALYSIS

Welcome to the requirement phase.
Where product vision turns into actionable plan.

1 **User Roles & Personas** Define real users to ensure features address genuine needs	2 **User Story Mapping** Visualize and end-to-end pet care journeys
3 **Initial Product Backlog** Conduct Sprint Zero to define and prioritize epics and features	4 **Prioritization Techniques** Rank items by value, risk, and dependency
5 **Roadmap** Map epic to sprints with outcome-driven themes	6 **MVP vs. MMP** Differentiate initial validation release from marketable product
7 **Backlog Refinement & Decomposition** Break down large, vague items into smaller, clear, and actionable user stories that are ready for upcoming sprints.	

SPRINT 0

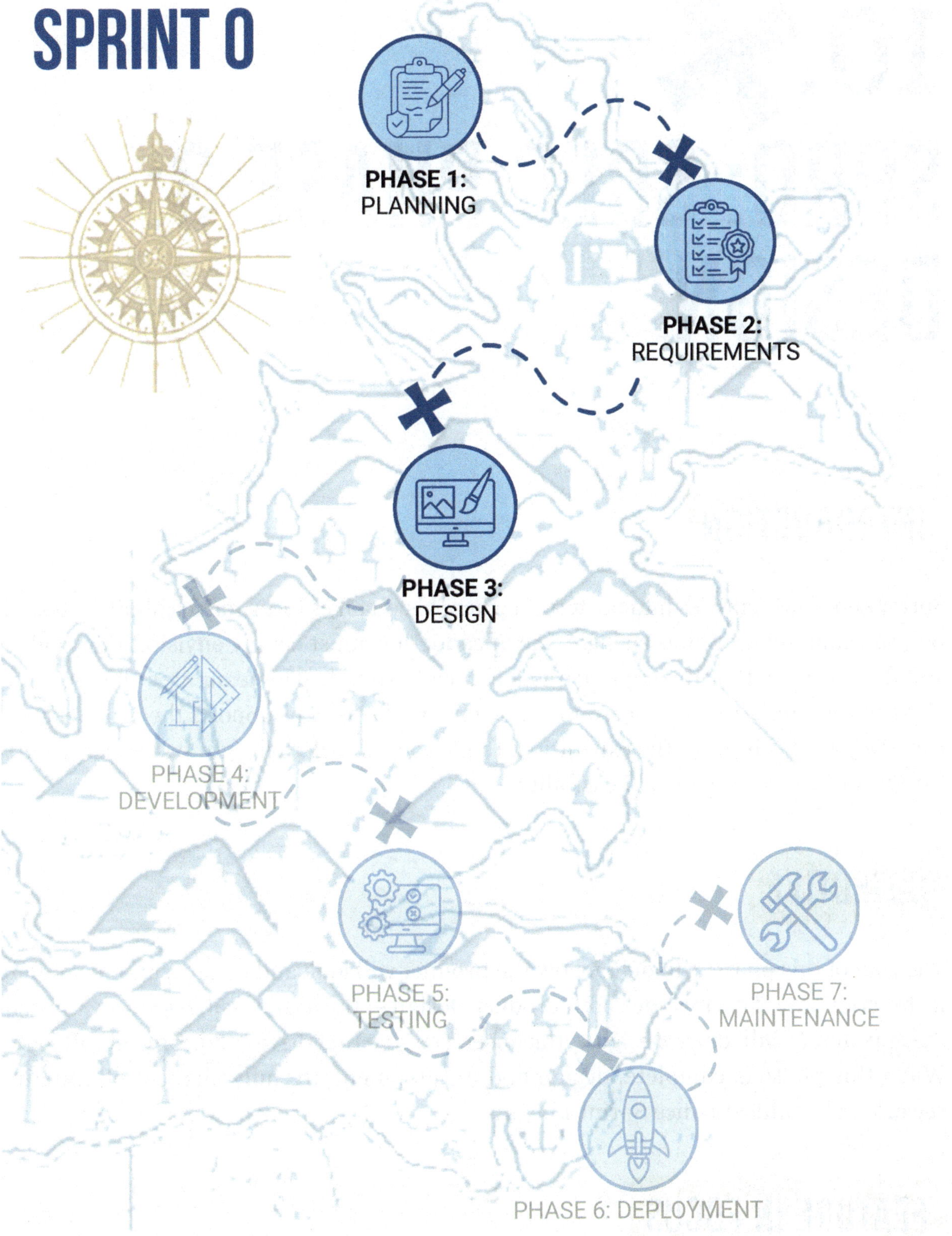
PHASE 1:
PLANNING
PHASE 2:
REQUIREMENTS
PHASE 3:
DESIGN
PHASE 4:
DEVELOPMENT
PHASE 5:
TESTING
PHASE 6: DEPLOYMENT
PHASE 7:
MAINTENANCE

10.

SPRO: SDLC PHASE 3: DESIGN

INTRODUCTION

Sprint Zero now ventures into the terrain where vision takes form and feasibility is tested. Before Team Vertex writes a single line of code, they must clearly envision the product and demonstrate that the approach is technically sound. The team uses this phase to align expectations, reduce uncertainty, and ensure the design supports the product goals from Day 1. At this leg of the expedition, the heroes unlock their first two superpowers: Design Vision and Technical Feasibility.

OBJECTIVE

The goal of this phase is to define how the product will look and behave, and to validate its functionality. This includes both Product Design — the user's visible journey through the app, and Architecture Design — the unseen engine that drives and connects all parts. When this phase is complete, design and development are fully aligned, eliminating rework and confusion once Sprint 1 begins.

FEATURE IN FOCUS

In Sprint Zero, the "feature" is alignment — a shared understanding that prevents wasted effort. This chapter focuses on two key areas: UX/UI Design, including wireframes, mockups, clickable prototypes, and architecture design, as well as Navigation, data flow, app structure, and risk validation. These serve as the blueprint for all future PetTracker features.

The team's design focus centers on establishing a clear and collaborative foundation for product development. They begin with wireframes, creating low-fidelity layouts that clarify the user interface and promote team alignment. From there, prototypes are built to simulate clickable flows, allowing early testing of user behavior and collection of meaningful feedback. The architecture map defines the structural plan, ensuring technical feasibility and guiding smooth implementation. Alongside this, a well-defined navigation flow outlines screen-to-screen logic, minimizing guesswork during development. Finally, risk spikes are used to conduct targeted technical validations, helping the team identify and prevent potential blockers before they impact progress. The adventure cannot begin until the map exists.

TEAM VERTEX IN ACTION

Jenna — UX Designer

She sketches the first screens: Pet Profile, Log Feeding, Log Walking, Log Medication and Home sceme.

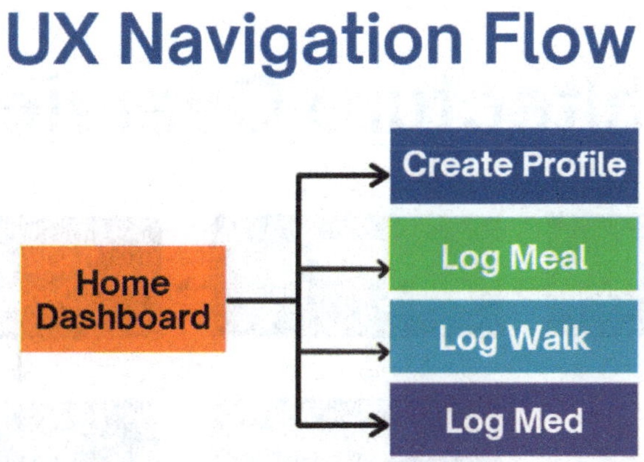

Exhibit 9.1 – Wireframe Navigation Map: Early visualization of core navigation between key PetTracker MVP screens to establish shared understanding before development.

She ensures accessibility standards and error states are included from the start. She doesn't work in isolation. From Day 1, she pulls in Alex and Malik to walk through her flows, identify confusion points, and lock down screen states and error paths.

Alex — Developer

He drafts the initial folder structure, sets up the navigation stack, and outlines where state and data will live.

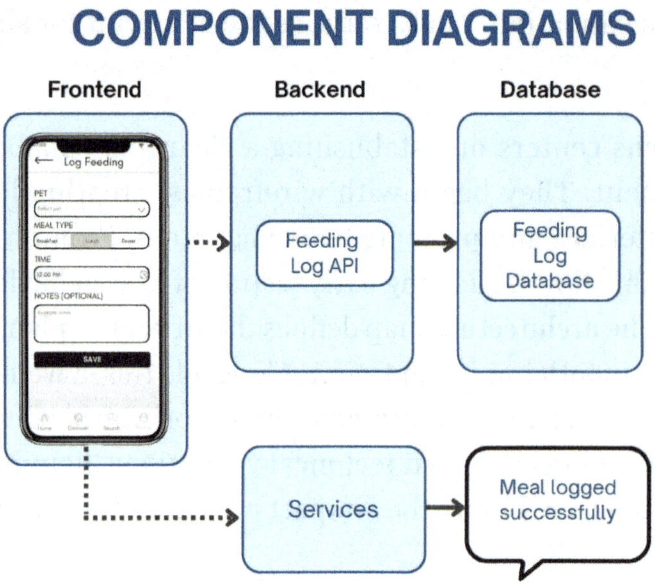

Exhibit 9.2 - Architecture Component Diagram — The Feeding Log feature connected from UI → API → Database with feedback returned to the user. This validates early feasibility and helps developers + testers align on how data flo ws through the system.

He runs a spike to confirm Expo + React Native support the dual-platform requirement.

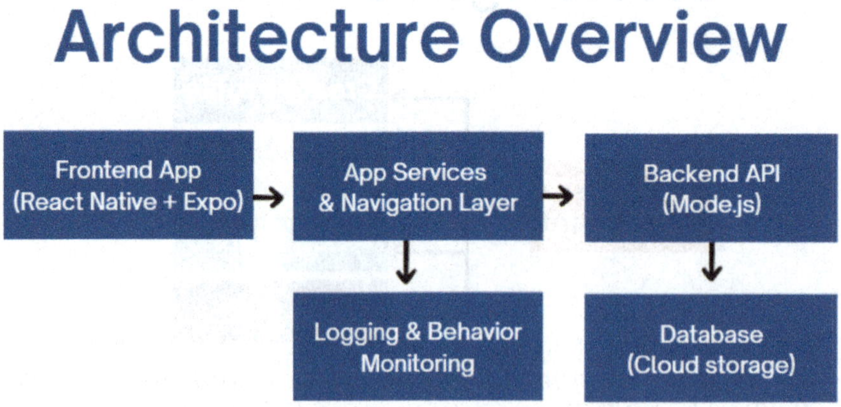

Exhibit 9.3 - Frontend-to-backend interaction model confirming structural feasibility and technical integration points for future feature delivery.

Malik — QA Analyst

He reviews prototypes for testability — can every behavior be validated?

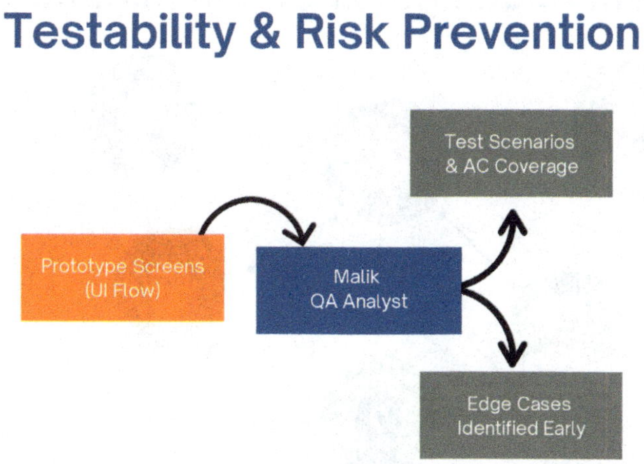

Exhibit 9.4 — QA Validation of Design Artifacts — Malik reviews early prototype screens to define test scenarios and identify edge cases well before development begins. He identifies early edge cases before they become expensive defects.

Maria — Product Owner

She ensures every design element ties directly to user value and business priority.

No extras. No over-designing. Only what matters.

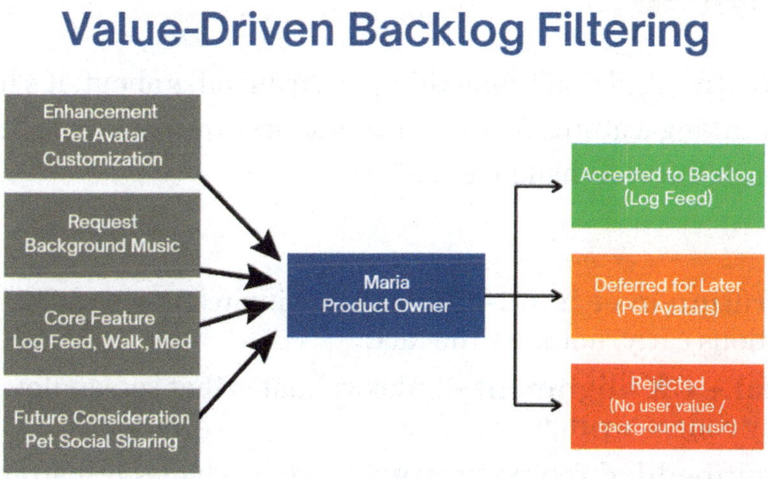

Exhibit 9.5 — Product Owner Value Filtering: Maria evaluates proposed ideas through a value lens — ensuring only high-impact features enter the product backlog, while lower-value requests are deferred or rejected entirely.

Maya — Scrum Master

She keeps alignment tight and communication flowing — no assumptions survive the day.

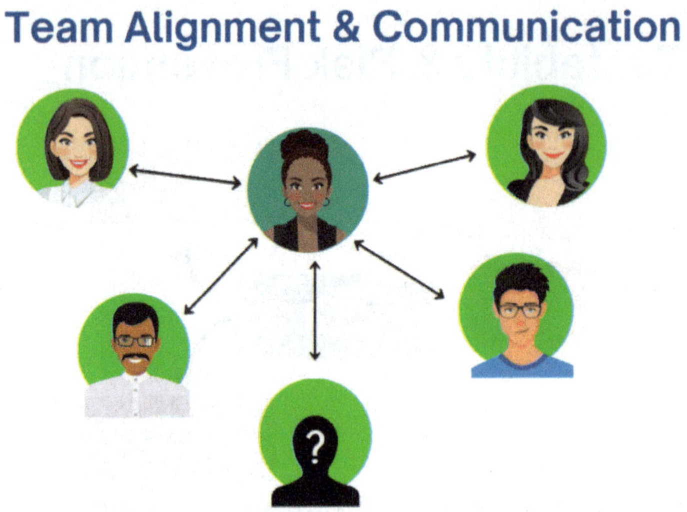

Exhibit 9.6 — Sprint Zero Alignment Flow: Maya enables continuous communication and rapid blocker surfacing so the whole team remains aligned and no assumption goes unvalidated. The team leaves this phase united by a clear visualization of the user experience and a confident architectural structure.

AGILE IN MOTION

Sprint Zero proves that Agile isn't something this team talks about; it's how they operate. Even without a working feature, Team Vertex practices the core Agile behaviors that will carry them through the real challenges ahead:

- **Collaboration over hand-offs** — Jenna invites Dev and QA into design conversations early, not after the fact.
- **Technical + UX alignment** — Alex validates that screen flows are feasible before any code is written.
- **Quality embedded from the start** — Malik shapes test strategy from prototypes, preventing costly rework later.
- **Lean product thinking** — Maria cuts feature noise and defends user value ruthlessly.
- **Team connectivity** — Maya ensures alignment is constant and communication is direct.

By the end of this phase, every team member shares the same mental model of the product. That alignment is the first increment — invisible to users, but essential for fast delivery. Sprint Zero isn't about velocity. It's about removing drag.

TEAM REFLECTION

Team Vertex pauses to assess how they worked together during the design process and how they want to improve going forward. They ask themselves: Do we all see the same product? Are we eliminating uncertainty early? Are we making decisions fast enough?

They conclude strongly, with two important improvements:

- **Design visibility upgrade:** Figma prototypes will now be embedded directly in every user story — no searching, no assumptions.
- **Risk-first planning mindset:** Any workflow ambiguity or unknown technology becomes a quick spike before it can become a roadblock.

The team leaves this reflection with confidence:

- They understand the user experience
- They trust the architecture they're about to build
- They see the same map for the journey ahead

They're not just "ready to start coding," they're **aligned, intentional, and dangerous to delay.**

DESIGN-DEV HANDOFF PLAN

Bridging Creativity and Code

Handoff Tools

Figma Inspect, Zeplin, Storybook, or
Dev Mode for spec access.

Asset Exporting

Fonts, icons, images, SVGs, etc.,
optimized for web/mobile.

Acceptance Criteria Alignment

Ensure designs correspond and linked to
the user stories in the backlog

Interaction Notes

Document Hover states, animations,
edge-case behavior and responsive behavior.

Version Control of Design Files

Use clear naming conventions and
shared "approved" folders.

In Agile, a smooth handoff between design and development is essential to avoid disruptions and ensure proper implementation.

TECHNICAL DESIGN & ARCHITECTURE SKETCHES

Building the Blueprint Before the Code

System Architecture Basics

Understand how your app is structured: frontend, backend, database and external services.

Component Diagrams

See how features like “Log Feeding” flow through the app, from UI to API to storage/database.

Data Flow & Sequence Diagrams

Track how information moves step-by-step and which parts of the system are involved.

Tech Stack Decisions

Learn how Team Vertex choose tools like React Native, Firebase, and GitHub to power PetTrackr.

Security Sketches

Keep user data safe with smart authentication, encrypted storage, and role-based access.

Takeaway: Good architecture doesn’t have to be complicated - it just needs to be clear. When you sketch before you code, you build with purpose and prevent confusion later.

SDLC PHASE 3: DESIGN

Welcome to the design phase.
Where ideas are shaped into tangible blueprints.

Wireframing

Sketching UI layouts and interactions

Mockups & Prototypes

Detailing colors, styling and flows

Technical Design Notes

Documenting technical design notes

Design-Dev Handoff Plan

Plan for a smooth handoff between design and development

Exercises & Activities

"Prototype to Alignment Check"

Applying UX + Feasibility + Testability from Day Zero

Scenario

You're stepping into Team Vertex as a Design + Delivery Contributor. The Product Owner just provided a new user story for the PetTracker app:

User Story:

As a pet parent, I want to edit my pet's profile so I can correct mistakes or update details over time.

Your mission: Ensure this story is fully visible, buildable, and testable before Sprint Planning.

Your Task

Follow this exact mini-workflow Team Vertex used in Sprint Zero:

1. Create a Low-Fidelity Wireframe

Sketch the Edit Pet Profile screen:

At minimum, include:

- Pet name field
- Pet type selector
- Birthday selector
- Photo upload/change button
- Save + Cancel actions
- Inline validation (e.g., "Pet name is required")

Can be drawn by hand, Miro, Figma, or even PowerPoint.

2. Define Behavioral Expectations

Write 2–3 Gherkin format acceptance criteria.

Example starter:

Given existing profile details loaded

When the user updates the name and taps save

Then the updated profile is stored and displayed on the Pet Profile screen

You expand from there.

3. Add One Feasibility Note (Dev Lens)

What data must already exist?

Where will the change be saved?

Do we need new API endpoints or are we updating existing ones?

4. Add One Testability Note (QA Lens)

What edge case would you validate?

(e.g., invalid date, oversized photo, empty required fields)

What to Submit

You should have 4 artifacts:

- Your wireframe sketch (photo is fine)
- Acceptance criteria (Given-When-Then)
- One feasibility note
- One edge case for testing

Upload them together — this becomes your first portfolio-ready UX/design artifact.

Why It Matters

You're designing screens and eliminating ambiguity before developers write a line of code.

That's how Sprint Zero saves teams from Sprint Headaches.

Level-Up Prompt (Optional)

Reflect:

Where did uncertainty still exist... and who could clarify it fastest?

This fosters true Scrum thinking — clarity is a team sport.

CHAPTER SUMMARY

Sprint Zero's design work gave the team something just as vital as code: clarity. Team Vertex established the visual language and user flow for the app long before a single feature entered development. Wireframes evolved into clickable prototypes, making user experience testable early, and ensuring that every interaction supported real user needs.

Design wasn't done in isolation. Jenna worked shoulder-to-shoulder with Alex and Malik so every screen was not only intuitive but buildable and testable. Architecture diagrams linked the UI to backend reality, eliminating guesswork and aligning feasibility with vision. Acceptance criteria flowed directly from these prototypes, giving developers and QA a shared truth to build from.

Before Sprint 1 even began, the team had a unified mental model of how the app should look, feel, and behave. That meant no pixel-perfect surprises later and no usability "revisions" disguising rework. Sprint Zero design didn't just show the product — it made sure the team saw the same product. This chapter locked in usability, cohesion, and confidence. Now the vision can translate cleanly into working software — without drifting off course as the adventure begins.

WHAT'S NEXT?

The vision is now visible, testable, and aligned. Team Vertex has a shared blueprint for the product and confidence that what they're building truly serves the user. But screens and diagrams don't ship to the App Store. Next, the team must transform design clarity into development capability — setting up environments, enforcing code standards, and proving that builds can run anywhere, anytime. Before launching into Sprint 1 delivery, they need a system that supports fast progress without friction.

Design unlocked the look and feel. Now the team unlocks the power to build.

Up next: *Sprint Zero — Development Enablement*

Where ideas gain muscle, pipelines come alive, and the quest transitions from planning... to action.

WIREFRAMES & PROTOTYPES THAT SPEAK USER

Wireframes translate user stories into visual design, clarifying the user interface before coding

WHY THIS STEP MATTERS

"Wireframes are where empathy meets structure—my job is to sketch what the user needs before a single pixel is polished."
— Jenna, UX Designer

LOW-FIDELITY WIREFRAME

CLICKABLE MOCKUPS

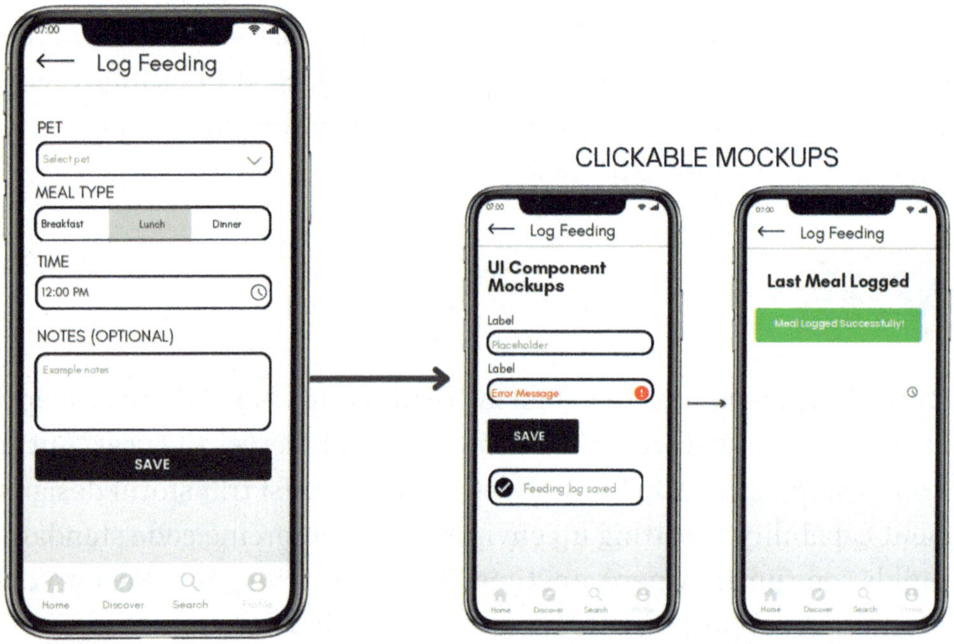

"The best wireframes aren't pixel-perfect, they're conversation starters."
— Maya, Scrum Master

WIREFRAME AS SHARED DOCUMENTATION

- Linked in Jira stories
- Used in design reviews
- Demoed in Sprint Reviews

SPRINT 0

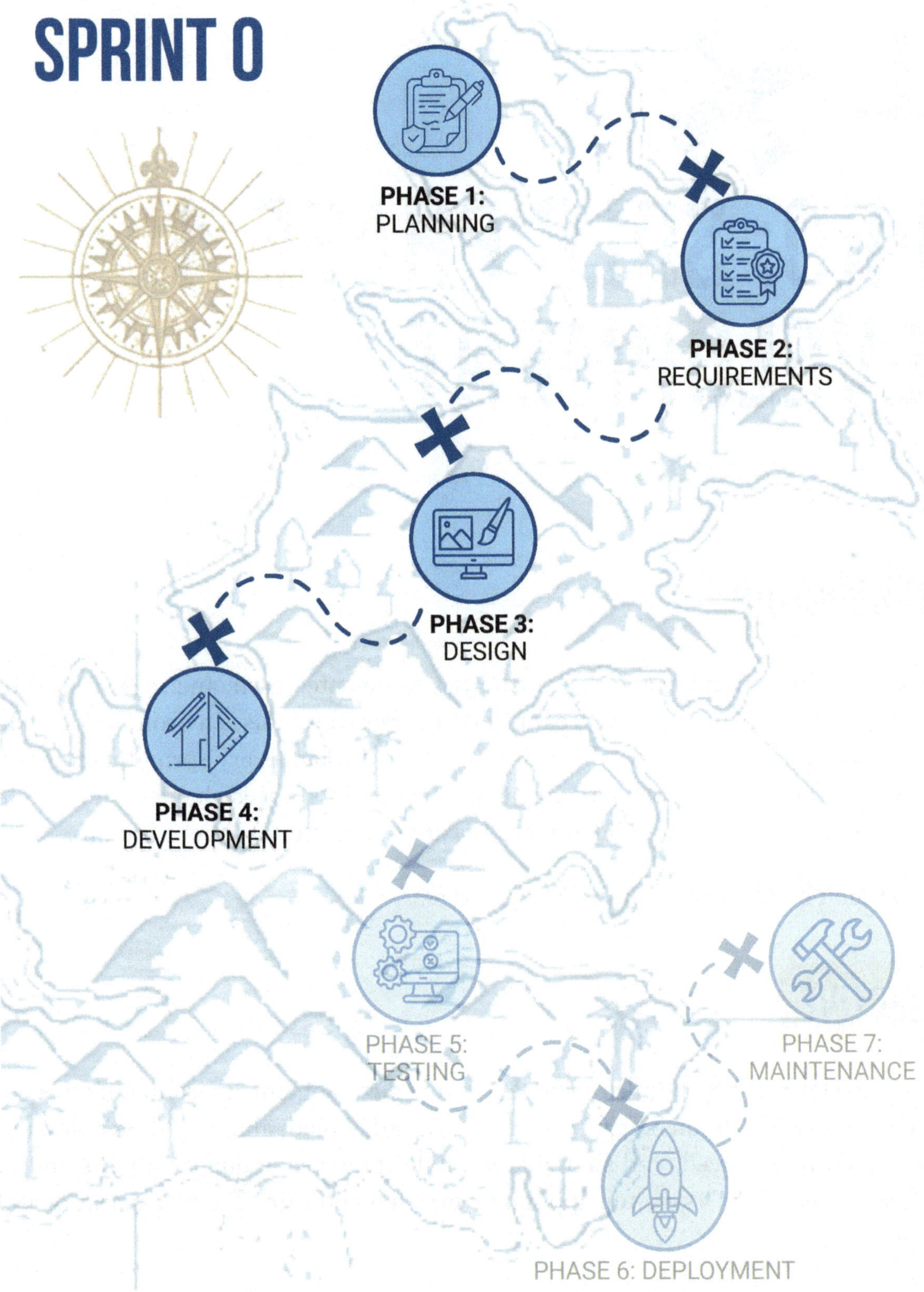
PHASE 1:
PLANNING
PHASE 2:
REQUIREMENTS
PHASE 3:
DESIGN
PHASE 4:
DEVELOPMENT
PHASE 5:
TESTING
PHASE 7:
MAINTENANCE
PHASE 6: DEPLOYMENT

11. SPRO: SDLC PHASE 4: DEVELOPMENT

INTRODUCTION

A quest is won by both ideas and the capabilities to execute, learn, and recover fast. In Sprint Zero's development phase, Team Vertex forges the tools, environments, code practices, and collaboration rules that will guide them through every Sprint that follows. This is where the "how" becomes undeniable. The mission of this chapter is to enable every developer and tester to build and run the app successfully without friction.

If anyone can't build locally, nothing else matters.

OBJECTIVE

Empower the team to begin Sprint 1 with a working development environment on every machine, repository + branching structure locked and loaded, automated PR rules and coding standards enforced, and shared accountability for code quality. By the end of this phase, Team Vertex's skeleton app runs everywhere emitting the first sparks of real delivery.

FEATURE IN FOCUS

The deliverable for this phase is the system that builds features. Your MVP won't get written if your repo is broken or your CI pipeline is flaky. The team's development capability is built on structure, quality, and collaboration. They start with a well-defined code repository structure, establishing Main, Develop, and Feature branches to support parallel work without code collisions. Strong standards and automation underpin their process through linting, code reviews, and pull request checks that prevent defects early and maintain consistent code quality. Ensuring local build readiness allows both Development and QA teams to run the app smoothly, fostering shared confidence in system stability. Finally, a clear collaboration flow, anchored in effective Git practices and pairing routines, helps eliminate lone-wolf bottlenecks and cultivates a culture of teamwork and collective ownership.

Constraints shape capability and rules unlock speed.

TEAM VERTEX IN ACTION

Alex and Malik rolled up their sleeves to prepare the development and testing environments. Before coding began, they worked side-by-side to install all necessary tools on their machines: Node.js, npm, the React Native framework, Expo for easy device testing, Android Studio, and Xcode for simulators, everything needed to run the PetTracker app locally. They cloned the empty project repository, ran the initial app scaffold, and verified it could launch on both an Android emulator and an iOS simulator without crashing. This hands-on pairing ensured that both the developer and QA perspectives were considered in setup: if Alex ran into any configuration snags, Malik was right there to help troubleshoot, and vice versa. By the end of this exercise, everyone on the team could build and run the app consistently. It was a simple skeleton app, but getting it running everywhere was a big confidence booster.

Here's a low-fi mockup visualizing Alex and Malik's paired environment setup for Sprint 0. It highlights the side-by-side laptops, installed tool stack, git clone step, and successful Android/iOS launches.

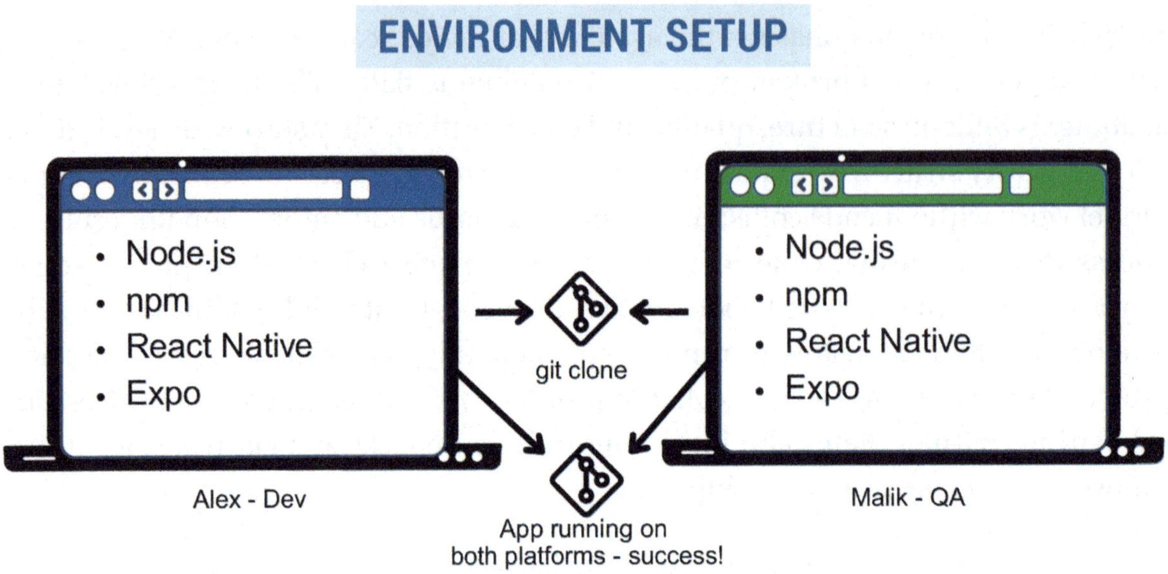

Maya coordinated the repository setup and version control strategy to enable smooth collaboration. She created a project space on GitHub for the team and set up the repositories for the mobile app and its backend. Together, they established a clear branching model: a permanent main branch for stable release-ready code, a develop branch for integrating ongoing work, and short-lived feature/* branches for each user story or task in development. The team also agreed that any urgent fixes after release would go into hotfix/* branches. With this strategy, Team Vertex could work in parallel without "stepping on each other's toes," as Maya put it. To enforce quality, they decided that all code changes would go through pull requests, and each PR must be reviewed by a teammate and pass automated checks (like build and lint tests) before merging. These agreements, made early, set a professional tone for how the team would write and share code.

Here's a wireframe showing Maya's branching model and pull-request workflow in GitHub: main, develop, feature/, and hotfix/ branches, plus the review + checks gate.

BRANCH MODELING & PULL REQUEST

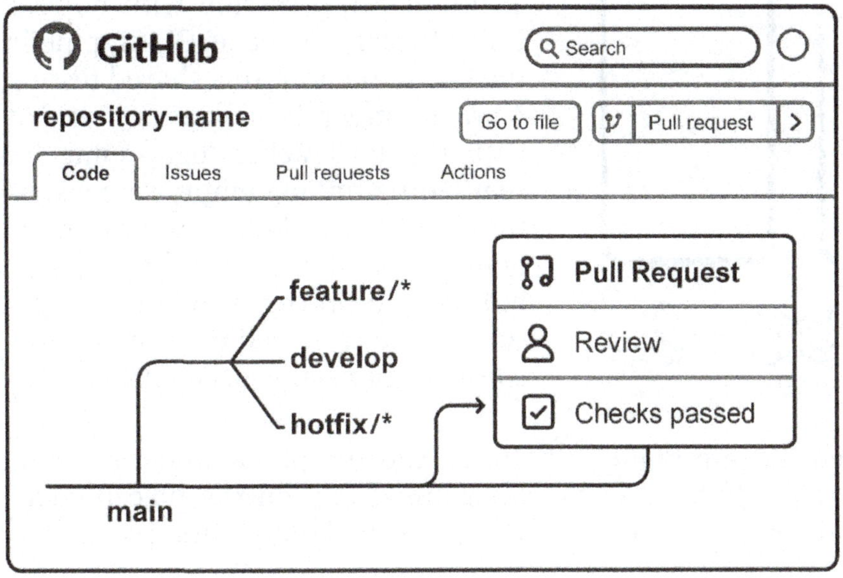

Throughout Sprint 0, communication was constant. Maya introduced Daily Scrum stand-ups even though the team wasn't delivering user stories yet. Every morning, they gathered briefly to share progress on setup tasks and surface any blockers. This helped catch issues early, for instance, on Day 2, Alex reported, "The Android emulator is giving me an error when launching the app," prompting a quick group fix that same day.

They treated these stand-ups as practice for the rhythm of real Sprints, building trust and transparency.

Maria (Product Owner) also facilitated a couple of backlog refinement sessions during this time, ensuring the initial user stories for Sprint 1 were clearly defined and prioritized. Together, they broke high-level epics into smaller user stories and checked that each story met their newly drafted Definition of Ready (e.g., each story had acceptance criteria, was estimated, had a design attached, and had no blockers). By the end of these sessions, the team had a Sprint 1 backlog that was groomed, pointed, and ready to go, another big win for alignment.

The climax of Sprint 0 came when Jenna and Alex collaborated to deploy the first build of PetTracker: a minimal app version with no real functionality. Using the freshly set-up CI/CD pipeline, they pushed the “skeleton app” to a shared testing environment. It wasn’t much to look at: a branded splash screen with the PetTracker logo, some basic navigation scaffolding linking empty screens, and placeholder text where real data would eventually go. But this deployment was a crucial milestone for the team’s morale and process validation. It proved that their source control, build scripts, and environment configurations all worked in concert.

Malik installed the app build on both an Android phone and an iPhone to verify that it launched successfully and that navigating between the placeholder screens didn’t produce any errors or crashes. Everything worked on the first try.

Team Vertex even held a Sprint Review (even with no real features) to demonstrate this foundational accomplishment to stakeholders. Maria (PO) set the context at the review meeting, reminding everyone that “This isn’t a feature demo: it’s a confidence check. We’ve proven our app can be built, shipped, and run on real devices. That clears the path for Sprint 1 to start strong.” In the live demo, Alex took attendees through the app’s placeholder interface, tapping through the dummy screens for Pet Profile, Log a Meal, Log a Walk, and Log Medication to show that the navigation flow was intact and the UI matched the intended design framework. Malik chimed in to confirm that he had tested the build on both iOS and Android, and all screens loaded quickly with no routing issues or unexpected crashes. There was nothing “usable” for an end user yet, but the stakeholders could already see the app’s structure taking shape and appreciated the polished look and responsiveness of the early UI.

One senior stakeholder commented on how reassuring it was to see tangible progress on the app setup. Maria did note a small improvement for the team’s backlog, adding header titles on each placeholder screen to help orient testers in the future, but overall, the review ended on a celebratory note. The outcome was clear: the team’s technical setup and design alignment were validated, and with this skeleton app in place, Sprint 1 could begin with confidence and momentum.

AGILE IN MOTION

Sprint Zero proved that Agile is about removing the friction that slows speed down later. Team Vertex demonstrated real delivery discipline before a single user story ever hit development:

- Working software early, even if minimal — the skeleton app validated that the pipeline, tools, and machines worked in harmony

- Automation before scale — the CI/CD pipeline became a non-negotiable teammate, not a technical afterthought
- Team-first ownership of quality — linting and PR checks ensured clean code with no hidden messes waiting to explode later
- Pairing as the default move — Dev + QA solving environment issues together reinforced shared accountability
- Structures that enable flow — branching strategy, review gates, and coding standards gave the team freedom within boundaries

This is the phase where Agile stops being theory and becomes a technical operating system.

TEAM REFLECTION

The team meets to inspect one truth: If setup slows us down now, it will slow us down ten times more later.

Their insights spark immediate improvement:

1. Tech Consistency Upgrade

Branch naming conventions are standardized and added to working agreements so merges remain smooth and predictable.

2. Better Surfaced Test Thinking

Dev + QA commit to a weekly architecture + test sync, shifting test plans even further left.

3. Design to Dev Handoff Timing

A Design Review step is added to the workflow to ensure no story enters development with unanswered questions or missing assets.

Their retrospective isn't for flow, clarity, and zero waste. Sprint Zero development taught them something powerful: Maintenance isn't what you do after the product ships, but how you treat your system from day one. Their engine is now tuned. Their working agreements are no longer assumptions but now commitments. The team is aligned, confident, and ready to deliver.

SDLC PHASE 4: DEVELOPMENT

Sprint Zero development isn't building features.
It's building the **system to build features** — fast.
If Setup is slow now, Execution will be 10x slower later.

Unified Local Development Environment

Every teammate - dev & QA must be able to:

- Install and configure required tools
- Clone the repo and run the skeleton app
- Validate Android _ iOS builds work consistently

Repository Setup + Branching Model

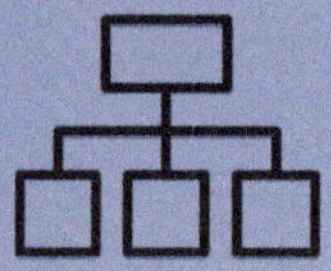

Foundation of scalable collabortion:

- GitHub repo created for app + backend
- Branch strategy defined; main, develop, feature
- Branch naming conventions documented

Automated Quality + Pull Request Workflow

Quality is baked in, not inspected later:

- All code changes go through PR reviews
- Build + lint checks auto-run before merge all
- Daily stand-ups reinforce transparency early

Paired Collaboration + Shared Ownership

No lone-wolf devs, no QA in the shadows:

- Dev + QA pairing on environment setup
- Peer review culture instead of "throw over the wall"
- Daily stand-ups reinforce transparency early

Skeleton App Deployment via CI/CD

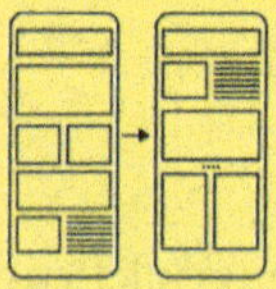

A tiny app - but a massive milestone:

- Branded splash screen + navigation setup scaffold deployed
- Runs on real devices without crashing
- First celebratory Sprint Review - even without features

Workflow & Readiness for Sprint 1

Structure that accelerates delivery:

- Sprint 1 backlog refined to Definition of Ready
- Acceptance criteria + pointing complete
- Design handoff checkpoints introduced

Exercises & Activities

Sprint Zero Development Exercise: "Build the Builders"

Your first step toward being production-ready

Scenario

The PetTracker skeleton app is about to come to life. But before anyone writes features, every team member must prove: The app builds, runs, and flows consistently. You're stepping into Team Vertex at the turning point between planning and execution. Your mission is to walk through the 3 key DevOps + Collaboration foundations that enable fast delivery later.

Your Mission (3-Part Challenge)

1. Branching Strategy Decoder (Collaboration Lens)

Here is Team Vertex's branching model:

- Main → stable releases only
- Develop → current working version
- Feature/ → each slice of work
- Hotfix/ → emergency production fixes

Your task: Match each situation below to the correct branch type:

Situation	Branch Used	Your Answer
A new story for Pet Profile screens		
A critical production crash affecting all users		
Integrating all Sprint work for demo		
App version submitted to App Store		

Ensures you understand where work flows.

2. Build Fail Triage (Technical Awareness Lens)

Malik runs the skeleton app... and it crashes. You see this error: Error: Image asset missing

→ logo.png not found. Which step below is the right Sprint Zero action?

Choose ONE:

A — Rewrite the entire app navigation

B — Add logo.png where the code expects it, rebuild, re-test

C — Log a bug and wait for Sprint 1 to fix it

D — Remove the splash screen altogether

This teaches fixing the system, not deferring problems.

3. Flow Confidence Check (Quality + Collaboration Lens)

One team member's laptop builds successfully. Another team member's laptop throws a permission error.

Your task:

Write one sentence that expresses the Sprint Zero mindset: How should the team respond?

Example starter: "If one person is blocked, the whole team stops to..." This reinforces shared accountability and zero-bottleneck culture.

You'll Produce a completed branching table, one correct build-failure decision, and one Sprint Zero "team rule" statement. Nothing in this exercise requires code, but everything in it protects future code from chaos.

WHY THIS EXERCISE MATTERS

Sprint Zero development ensures:

- No environment excuses
- No "works on my machine" nonsense
- No slowdowns once the real race begins

Because... Speed is earned by eliminating resistance before you run. You're not just preparing to build an app, you're preparing to build like a professional team.

CHAPTER SUMMARY

Sprint Zero didn't ship a visible feature. It built the engine that will ship all of them. Team Vertex leaves this phase with a product that may not "do" much yet, but a team and system that can now do everything they need to move fast without falling apart. Every developer and tester can build and run the app locally. The branching strategy and pull request rules create order before chaos has a chance. Automated checks protect quality without adding ceremony. Collaboration is baked into their tools, habits, and agreements.

The result is a working skeleton of the PetTracker app moving cleanly through a functioning CI/CD pipeline, proving that delivery is now operational, not theoretical. Sprint Zero has turned uncertainty into capability. It has transformed a concept into a reliable system that can ship value Sprint after Sprint, without getting stuck on setup, misalignment, or avoidable technical debt.

Team Vertex steps into Sprint 1 with momentum, confidence, and a development engine designed to scale. There's no polish for users yet, but there is structural readiness for the team. And in long-term product work, that's the real competitive advantage.

WHAT'S NEXT?

With the foundation laid in Sprint 0, Team Vertex is ready to hit the ground running in Sprint 1. In the next chapter, we'll see the team tackle their first real backlog items, starting to implement features like user authentication (Login) and the Pet Profile creation. The planning and preparation done in Sprint 0 will be put to the test as the team moves into execution mode.

How will their processes and plans hold up when building a feature for the first time? We'll follow Team Vertex through Sprint 1's full SDLC cycle, from planning and requirements to development and testing, all the way to a potentially shippable product increment. With a clear goal and a confident team, it's time to turn all that preparation into tangible progress on the PetTracker app. Sprint 1 awaits: the real work begins now, backed by the strategy and structure forged in Sprint 0.

SPRINT 0

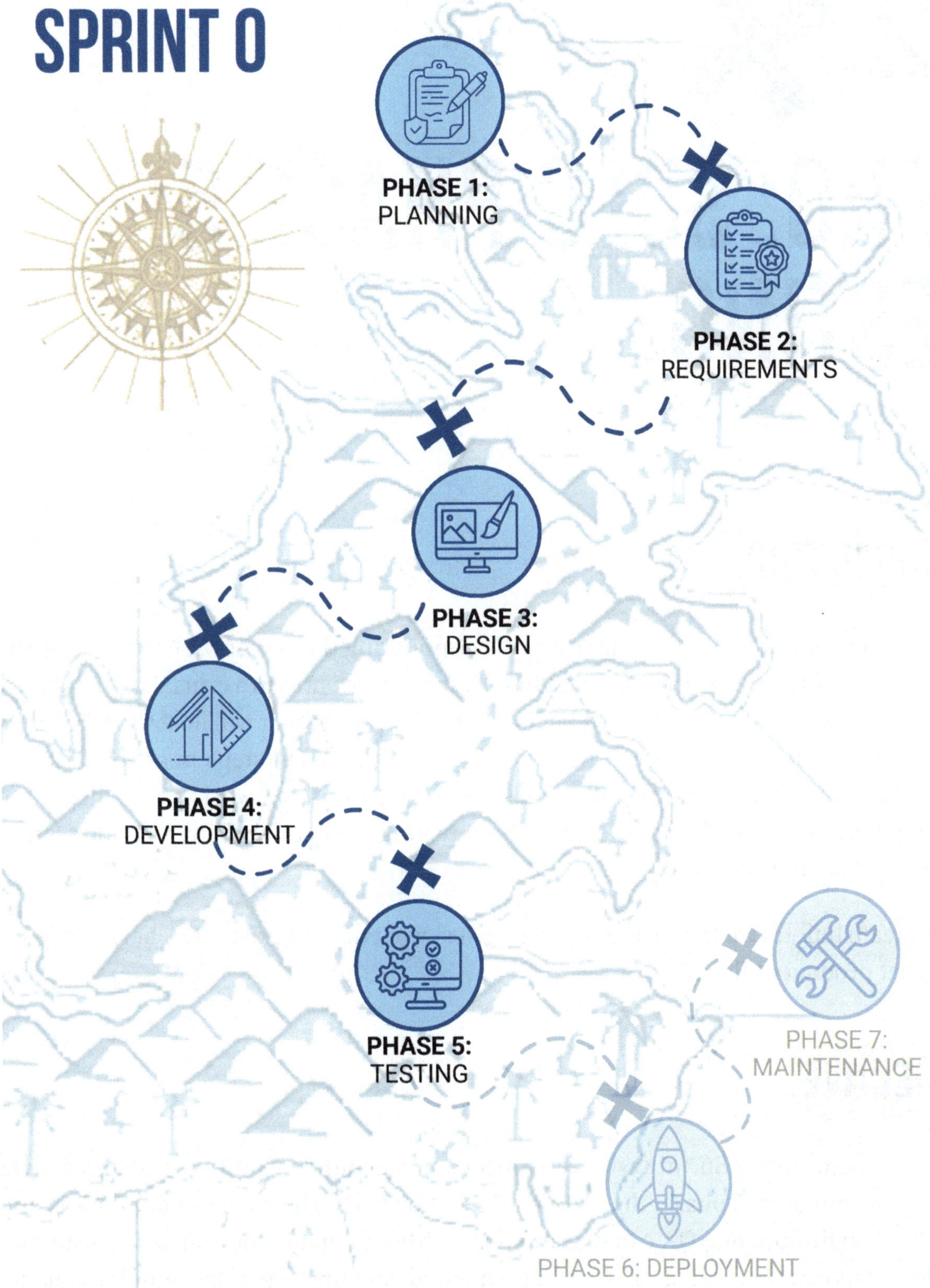
PHASE 1:
PLANNING
PHASE 2:
REQUIREMENTS
PHASE 3:
DESIGN
PHASE 4:
DEVELOPMENT
PHASE 5:
TESTING
PHASE 6: DEPLOYMENT
PHASE 7:
MAINTENANCE

12.

SPRO: SDLC PHASE 5: TESTING

INTRODUCTION

Bravery isn't charging into the unknown; it's eliminating the unknown before it has the chance to bite. In Sprint Zero's Testing Phase, Team Vertex proves a crucial truth: quality doesn't begin when features arrive; it begins when the system that will host those features comes to life. Before any end-user functionality is shipped, QA steps in as the first line of defense, validating environments, pipelines, navigation, and workflows. This phase isn't about testing individual stories; it's about testing the team's ability to build, verify, and trust what comes next.

Can the team build? Can they deploy? Can they test what doesn't exist yet? That confidence is the real product of Sprint 0.

OBJECTIVE

Ensure the team enters Sprint 1 with zero uncertainty about the ability to test and deliver fast. By the end of this phase, the following must be true: The skeleton app builds and runs in all environments, QA can execute tests without dependency on Dev, Test cases and checklists exist before features are implemented, and the CI pipeline enforces quality on every commit. No chaos, surprises, or "we'll test later." Later is when defects multiply and trust dies.

FEATURE IN FOCUS: FEEDING ACTIVITY

The "feature" of this phase is testing readiness, which is basically the ability to validate value before it impacts users. Core testing assets were prepared in Sprint Zero. The testing assets are designed to ensure reliability, consistency, and seamless collaboration throughout the development lifecycle. Smoke tests serve as the first checkpoint, confirming that the app opens and navigates correctly to validate overall build integrity. Environment tests enable QA teams to replicate development builds locally and across devices, preventing dependency blockers and saving time during validation. Standardized test case templates provide a uniform execution format that eliminates guesswork and ensures comprehensive coverage throughout the process. Clear defect logging rules establish a shared triage language between teams, reducing friction during issue escalation and resolution. Finally, setting up API tools, such as a dedicated Postman workspace, empowers testers to validate backend and UI components independently, strengthening the product's overall testability and stability.

If testing isn't ready, delivery isn't real.

TEAM VERTEX IN ACTION

Malik leads the charge as The Quality Guardian, but he isn't a lone ranger. He tests the skeleton app across devices, validates environment setup from a QA lens, confirms navigation flows match design, and ensures defects found now never return later

Alex also set up the team's first unit tests — short, focused checks to verify that key logic (like Firebase email handling and profile doc creation) works as expected. These tests run automatically in CI on every commit. The team agreed: if even one unit test fails, no code merges.

When the Android emulator threw a deployment error, he paired with Alex, solved it, automated a fix, and then documented the steps so nobody could trip on that same stone twice. That is shift-left testing: Find and fix it early. Then forget it forever. He also prepares the first risk-based smoke tests:

- App installs successfully? ☑

- Splash screen loads? ☑
- Navigation works between placeholder screens? ☑
- No crashes during tapping and swiping? ☑

This "Sprint 0 Smoke Suite" wasn't about proving features worked — it was about proving the system could hold weight. Malik tested on both Android and iOS, emulators and real devices, validating startup speed, UI transitions, and basic navigation. This suite became Team Vertex's baseline — the test every future build must pass before it's shown in a Sprint Review.

Smoke Testing = Confidence Check

"Is the app buildable? Launchable? Navigable? Crash-free?"

If yes → we move forward. If no → we fix now, not later.

He even logs the team's first "defect" — missing header titles — not as a mistake, but as a reminder that clarity is quality. Every win here pays interest in Sprints ahead.

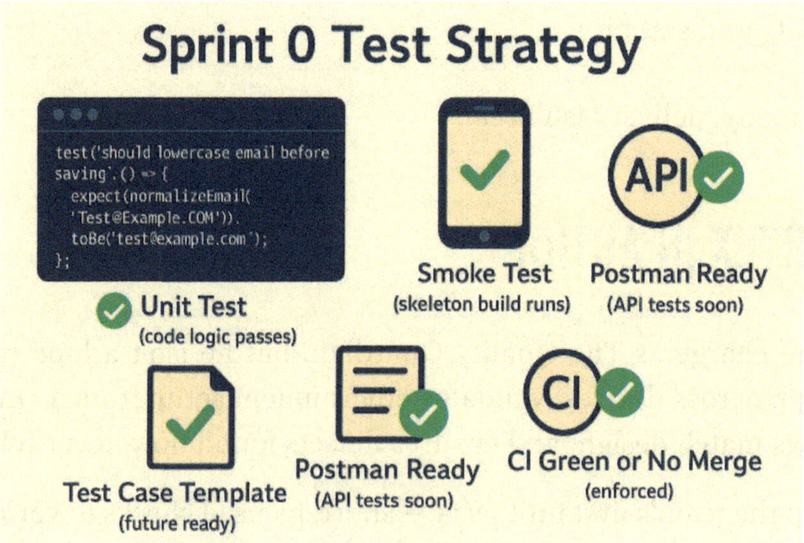

This infographic visualizes the testing foundation laid in Sprint 0, before a single feature is released. It highlights Team Vertex's key quality assurance practices to ensure that deployment, testing, and debugging workflows are ready before Sprint 1 begins. From smoke tests that validate basic app functionality (such as launch and navigation) to environment tests that ensure QA can operate independently, this map illustrates how testing readiness becomes a team-wide responsibility. Each icon represents a proactive safeguard: testing tools, automation triggers, and validation gates that prevent surprises down the road. In short, this is testing code and confidence.

AGILE IN MOTION

Team Vertex proves that Agile testing is a mindset, not a milestone.

Key Agile testing patterns established:

- QA joins design... before a ticket exists
- Developers expect and welcome immediate testing
- Defects become conversations, not accusations
- Test results drive backlog refinement
- "Working software" means working for everyone

Testing is shared accountability.

TEAM REFLECTION

The team gathers for a tight retrospective on testing capability: Did testing slow anyone down? Can QA work without Dev breathing down their neck? Are we catching risk long before failure reaches users?

Their improvements evolve into new agreements:

- ☑ Every story must include test notes — no vague outcomes
- ☑ QA access to any tool = instant priority
- ☑ No story moves to 'Done' without validation on both mobile platforms

Their takeaway: "Quality is something we design, enforce, and protect." Testing became the dragon shield for the adventure ahead. Sprint Zero didn't deliver user features. It delivered trust. The team walked away knowing their app could launch without drama, tests ran before code merged, bugs were treated as team opportunities, not individual blame, and testing was a mindset.

This phase wasn't called "Testing" because of QA. It was called that because this is where Team Vertex tested their system, their process, and their discipline.

SDLC PHASE 5: TESTING

The phase that focuses on catching bugs early and validating each feature

Purpose & Outcomes

Embed Quality Shift from "test after code" to "test with code", so defects surface immediately.

Validate User Value Confirm each story meets its acceptance criteria

Prepare for UAT & PROD Release Deliver a thoroughly tested, low-risk increment ready for stakeholder review and deploy

Testing Activities

Unit Testing	UI Testing	Functional Testing	Regression Testing	Test Automation
During Development	Mid-Sprint	Before Sprint Review	On-going	Once critical paths stabilize

Exercises & Activities

Exercise: The "Testing Readiness Audit" Think Like a QA

Scenario:

You're Malik, the QA for Team Vertex. Developers say they're ready to build in Sprint 1. But you're not checking code, you're checking preparedness. Your job is to audit your team's readiness to test.

Instructions:

1. Grab a pen and paper (or open a blank doc).
2. Imagine your team is about to deliver their first working feature in Sprint 1.
3. Answer the following testing readiness prompts (no tools needed):

Testing Readiness Prompts:

1. ***Environment Check***
    - Do you know where testing will happen? (e.g., "Is there a UAT or QA environment set up yet?")
    - Who has access? Is it shared or siloed?
2. ***Team Roles***
    - Do we know who will test during Sprints? (Just QA? Or Devs too? Will PO validate stories?)
3. ***Test Scenarios***
    - Without writing formal test cases, list 3 example "happy path" scenarios for the Login feature. (Ex: User signs up → logs in → sees dashboard)
4. ***Acceptance Criteria Readiness***
    - Pick one story (like "As a user, I can log in"). Write what a good "Definition of Done" for that story might look like. How do you know it's "done"?

5. ***Bug Plan***

- If someone finds a bug, what happens? Where do you report it? What info should you include?

Deliverable:

Write a short "QA Kickoff Memo" answering the 5 prompt areas above. Format:

- Environment:
- Team Testing Roles:
- Example Scenarios:
- Definition of Done Sample:
- Bug Reporting Plan:

Turn it in like you're prepping the team before Sprint 1.

CHAPTER SUMMARY

Sprint Zero Testing didn't validate finished features; it validated the system of delivery. Team Vertex confirmed that the app can be built and launched consistently, that QA can work without waiting on ad-hoc support, that navigation and environments are stable enough to trust, and that defects discovered early become shared improvements rather than production fires.

By pushing testing left into environments, pipelines, and basic flows, they reduced the biggest hidden risk in Agile: discovering foundational problems only after stakeholders are watching. No user saw any of this work, but the team walked away with something priceless: confidence. Confidence that Sprint 1 won't crumble under missing setups. Confidence that every increment can be verified quickly. Confidence that quality is built-in, not bolted-on.

Team Vertex ends this phase aligned around one belief: quality isn't a department; it's how they build. The heroes are ready. The world just hasn't seen the app yet.

WHAT'S NEXT?

Team Vertex now knows the product can be tested. Next, they must prove the product can be delivered. Sprint Zero isn't complete until every increment, even a placeholder one, can be deployed on demand: code builds automatically, pjpelines run without babysitting, the app installs cleanly on real devices, and every environment is ready for action. The Testing phase removed uncertainty about quality. The Deployment phase removes uncertainty about delivery. The question now becomes: Can we ship what we build whenever we choose to? Because the moment Sprint 1 ends, the team must show working software to stakeholders.

Next up: Sprint Zero — Deployment & DevOps Enablement

Where "Done" evolves from a definition into a capability.

SPRINT 0

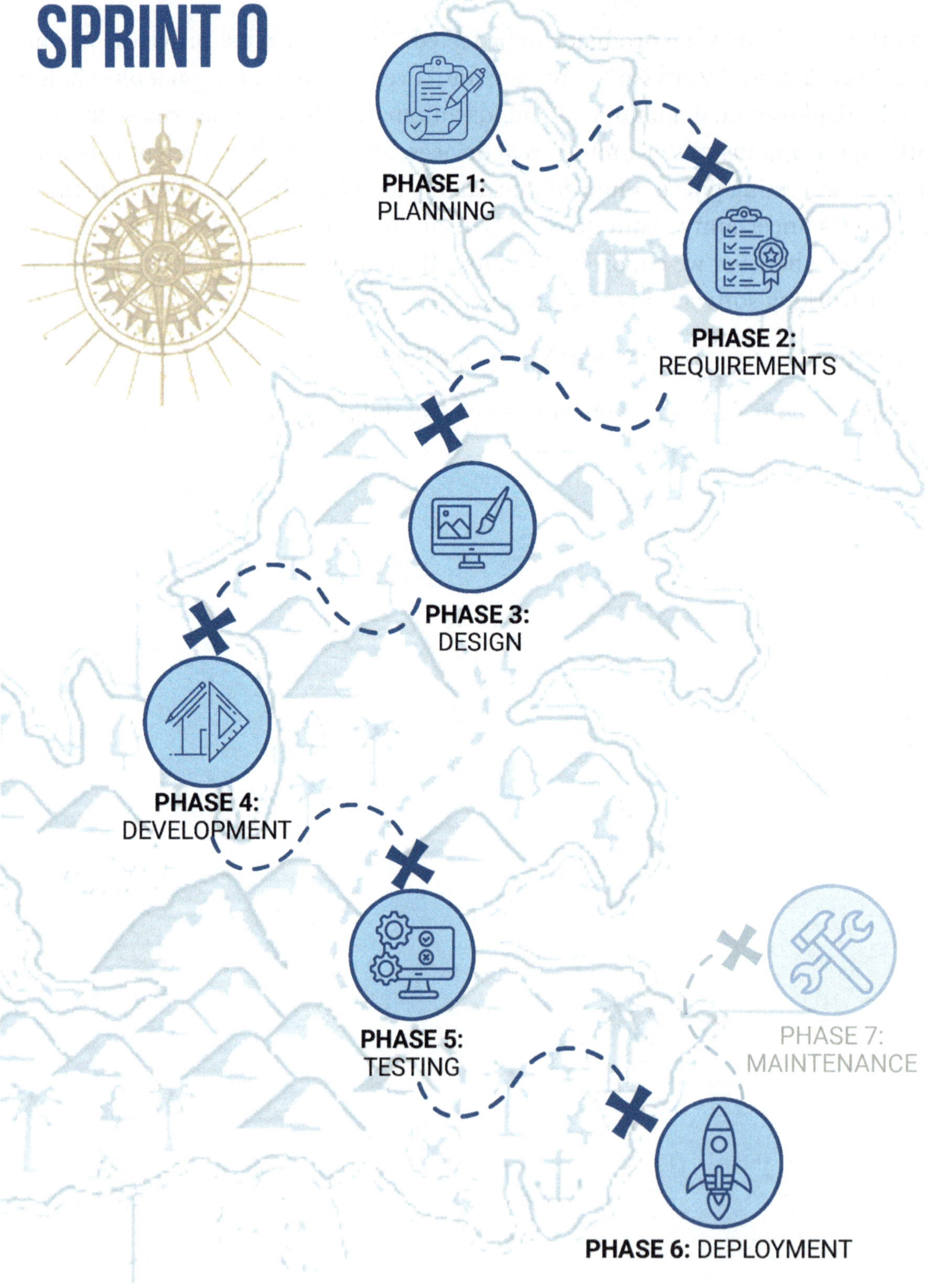
PHASE 1: PLANNING
PHASE 2: REQUIREMENTS
PHASE 3: DESIGN
PHASE 4: DEVELOPMENT
PHASE 5: TESTING
PHASE 6: DEPLOYMENT
PHASE 7: MAINTENANCE

13. SPRO: SDLC PHASE 6 DEPLOYMENT & DELIVERY CAPABILITIES

INTRODUCTION

A product that can't be delivered is a prototype with ambition. In Sprint Zero's deployment phase, Team Vertex wires up the CI/CD pipeline, pushes the first skeleton build, and proves end-to-end delivery in the real world. This is a confidence chapter: builds are reproducible, installs are predictable, and stakeholders can watch an app boot on a device without crossed fingers.

OBJECTIVE

The goal of this phase is to establish a minimal yet reliable delivery system that will enable rapid iteration from Sprint 1 onward. By its completion, the continuous integration (CI) pipeline should automatically trigger on every commit to the development branch, running essential pull request checks such as build and lint validations to ensure code integrity. A skeleton build will be generated and deployed to a shared testing environment, whether through an internal web preview, Expo Go, TestFlight, or another internal track—providing a tangible artifact for early feedback. This setup guarantees that a demoable version of the product is always available for the

Sprint Review, reinforcing transparency and progress visibility. Finally, both rollback and re-run operations will be simplified to a single command, ensuring that recovery or redeployment is fast, predictable, and never a horror story.

FEATURE IN FOCUS

The "feature" is Deployment Capability—a working assembly line for shipping. The delivery capability in Sprint Zero focuses on automation, traceability, and stability. The CI workflow is set up to automatically build on pull requests to the development branch, catching potential breakages before code is merged. Establishing minimal quality gates—including linting, type checking, compilation, and bundling—helps prevent late-stage surprises and maintains consistency across the pipeline. A defined delivery target ensures that a functional skeleton of the app is deployed to a shared environment, giving stakeholders something tangible to review and interact with early on. Implementing versioning discipline through semantic versioning, build numbers, and a changelog stub keeps every release traceable and well-documented. Lastly, being rollback-ready with documented redeploy and rollback steps minimizes risk, ensuring that any deployment issue can be swiftly contained and resolved without major disruption.

Rule of thumb: if a new dev can kick off a build and get an installable artifact in under 10 minutes, you're Sprint-1 ready.

TEAM VERTEX IN ACTION

Maya (Scrum Master) convened a short "Deployment Readiness" working session to define the promotion path: DEV → QA → (UAT later) → PROD. For Sprint Zero, they kept it simple—single target: a shared test build accessible to the whole team and core stakeholders.

Alex (Dev) added a lightweight CI workflow that runs on PRs to develop: installs dependencies, performs sanity checks (bundle and lint), and packages artifacts for the skeleton. He also set required checks on development so nothing merges red.

Malik (QA) wrote a Deployment Smoke Checklist:

- App installs/loads without errors.
- Navigation across 4 skeleton screens succeeds.

- Startup time under a sensible bar (e.g., < 3s web preview, < 5s device cold start).
- No console/runtime red flags.

Maria (PO) prepared the Confidence Review, including a short demo script and acceptance notes. Outcome she wants: stakeholders leave saying, "We can ship increments weekly starting Sprint 1."

Jenna (UX) verified the visual entry point (splash/title/placeholder labels) matches what stakeholders expect to see, so value conversations don't derail into "what am I looking at?"

THE DRY RUN:

- Commit lands on develop → CI runs → artifact is produced.
- Malik installs/opens → validates "Launch OK / Navigate OK / No crash".
- Maya schedules a 10-minute Confidence Check with stakeholders. Alex drives the click-through; Maria frames what this unlocks next week.

AGILE IN MOTION

- Done means demoable: The team treats "delivery system running" as a real increment.
- Automate the boring: If a human repeats a step twice, it goes into CI.
- Protect the trunk: PR checks and branch protections reduce drama.
- Prove reversibility: A rollback plan isn't theater; it's tested at least once.
- Make lead time visible: Build duration + time-to-first-install become tracked metrics from day one.

TEAM REFLECTION

What worked

- The "single simple target" made Sprint-Zero delivery achievable in days, not weeks.
- Required checks on develop immediately raised code hygiene without slowing the team.

What to improve next

- Add a basic smoke test to CI (boot + route test for web, or a minimal unit test for RN).
- Introduce release notes template and build labeling (e.g., so-<date>-<shortSHA>).
- Capture install success rate and startup time as CI/CD dashboard widgets (even manual for now).

New agreements

- No PR merges if CI is red—no exceptions.
- Every Sprint Review includes a fresh build link/QR in the invite.

SPRINT ZERO - SDLC PHASE 6: DEPLOYMENT

Deployment Key Practices During Sprint Zero

Each build results in a working, installable version that stakeholders can see, touch, and test

Repetitive steps are automated through CI/CD pipelines, reducing human error and accelerating feedback loops.

Every build, commit, and version is tagged, logged, and documented, ensuring full visibility across the delivery pipeline.

A single command can revert the system to a stable state, minimizing downtime and making recovery swift and predictable.

Builds are consistent, predictable, and deployable at any time; reinforcing confidence that “Done” truly means “ready to ship.”

Exercises & Activities

"Delivery Readiness Dry Run" Simulation

Purpose: Reinforce the mindset that delivery is not a future problem—it's a Sprint Zero deliverable. Students will simulate what "delivery readiness" looks like, even if they're not wiring up CI/CD themselves.

SETUP

Scenario:

You're on Team Vertex. The Login skeleton feature is coded. Now it's time to simulate a real delivery pipeline using only index cards, sticky notes, or a Miro/Mural board.

Your goal:

Map and stress-test your team's delivery system before Sprint 1 begins.

PART 1: DELIVERY PIPELINE MAPPING

Create a simple visual (paper or Miro) that includes these steps:

1. ✅ Commit Code → Push to Develop Branch
2. 🔄 CI Trigger → Run Lint + Build Checks
3. 📦 Build Artifact → Expo / Web Preview / TestFlight
4. 📲 Malik (QA) Installs & Validates
5. 👥 Maria (PO) Prepares Confidence Demo
6. 💬 Stakeholders Review & Approve
7. ↩️ Rollback (if needed)

Use sticky notes to represent:

- Team members' roles
- Artifacts created (builds, checklists, rollback docs)
- Risks or "what could go wrong" points

PART 2: DRY RUN SIMULATION

With a partner or group, assign roles:

- Dev (Alex)
- QA (Malik)
- PO (Maria)
- SM (Maya)
- Stakeholder

Then simulate this dry run aloud or in writing: "Alex pushes a commit...what happens next? What if the build fails? What does Malik check for? What does Maria need to prepare?"

USE A CHECKLIST LIKE:

- ☐ Build succeeded?
- ☐ Artifact created?
- ☐ QA install tested?
- ☐ Demo script ready?
- ☐ Rollback plan validated?

ENCOURAGE IMAGINATIVE BLOCKERS:

- What if the Expo link expired?
- What if the build works on web but not mobile?
- What if the demo artifact is blank?

DELIVERABLE

- A visual pipeline map with team roles and steps
- A filled-out "Sprint Zero Confidence Checklist" (template provided)
- A 3-paragraph summary titled: "How We Proved We're Sprint-1 Ready (Even Without Code)"

CHAPTER SUMMARY

Deployment in Sprint Zero is the ultimate bluff-caller. It asks: Can you deliver, or are you just rehearsing? Team Vertex answered that question with action, not theory. They built the very first delivery pipeline, shipped the first skeleton build, and demonstrated that the app can transition from code to a live device while stakeholders observe. This wasn't a "look at our screens" moment. It was a battle test of the entire delivery system: source control discipline, build automation, artifact packaging, installation, smoke testing, and live demonstration.

With this step, Team Vertex established their shipping rhythm – small increments, frequent delivery, zero heroics. The pipeline is now a silent team member, enforcing code quality and catching problems early before they poison later Sprints. Stakeholders now have more than promises; they've seen evidence: a real app running on real devices.

This chapter ends with a critical transformation: Team Vertex no longer intends to deliver. They are a team that delivers. Sprint Zero deployment produced trust, confidence, and a repeatable path to production.

That is the foundation every future Sprint will stand on.

WHAT'S NEXT?

The pipeline works. The app installs. The path to production is real. Now the team must shift to the most underestimated phase of Sprint Zero:

Maintenance & Retrospective – the discipline of staying fast

In this final setup phase, Team Vertex will review every tool, workflow, and decision made in Sprint Zero, identify friction points before they harden into future slowdowns, implement proactive monitoring and maintenance habits, and create technical debt guardrails that protect quality long-term. This isn't about fixing bugs since the app barely exists. It's about removing process bugs before they become expensive. Team Vertex now asks: "How do we ensure this speed and clarity continue when real features hit?" Their answer becomes the final safeguard:

- Lightweight monitoring in place
- Branch discipline automated
- Testing visibility elevated
- Release rituals repeatable
- Ownership shared across roles

Once the team safeguards consistency, they earn the right to begin Sprint 1.

SPRINT 0

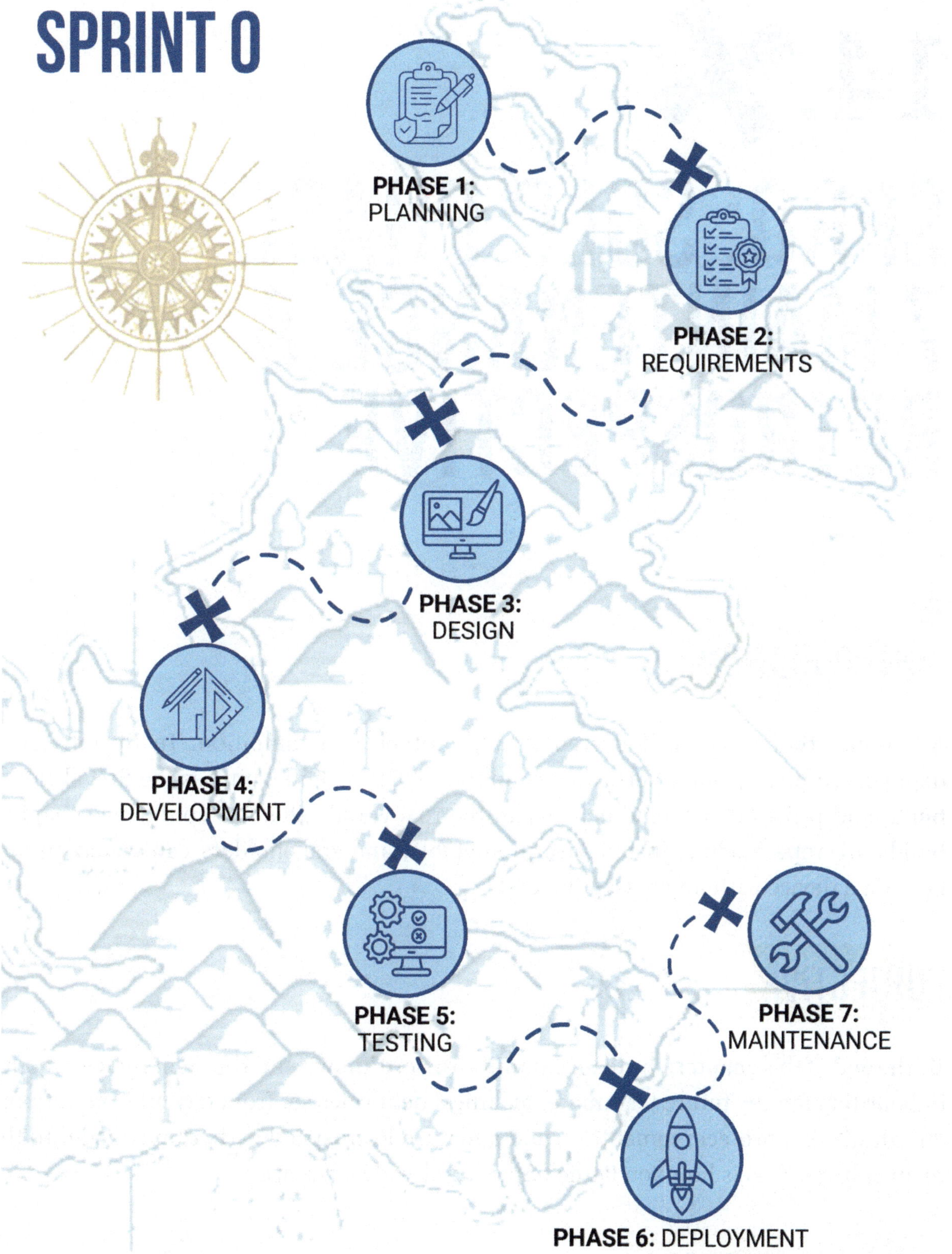
PHASE 1:
PLANNING
PHASE 2:
REQUIREMENTS
PHASE 3:
DESIGN
PHASE 4:
DEVELOPMENT
PHASE 5:
TESTING
PHASE 6: DEPLOYMENT
PHASE 7:
MAINTENANCE

14. SPRO: SDLC PHASE 7 MAINTENANCE & RETROSPECTIVE

INTRODUCTION

A product that can't be delivered is a prototype with ambition. In Sprint Zero's deployment phase, Team Vertex wires up the CI/CD pipeline, pushes the first skeleton build, and proves end-to-end delivery in the real world. This is a confidence chapter: builds are reproducible, installs are predictable, and stakeholders can watch an app boot on a device without crossed fingers.

OBJECTIVE

By the end of this chapter, the team commits to: Zero known environment issues, smooth onboarding for any future teammate, technical debt rules set from day one, operational guardrails that protect momentum, and a mature Retrospective rhythm baked into the team culture. This is how you launch Sprint 1 already optimized.

FEATURE IN FOCUS

There's no feature here. The "increment" is performance and flow assurance. This phase focuses on strengthening the foundation for long-term operational stability and continuous improvement. Operational health is maintained through a monitoring checklist and release verification steps, ensuring that mid-sprint outages are prevented before they occur. Flow protection is achieved by enforcing branch rules and automating pull requests, creating a development environment that flows smoothly rather than being a bottleneck. To sustain progress, process discipline is established through regular retrospectives and structured improvement cadences that drive continuous evolution. Meanwhile, knowledge capture is prioritized with clear README files and onboarding documentation, ensuring new contributors experience zero downtime when joining the project. In this phase, ⊘ no new functionality is added—instead, ✓ future functionality is protected, setting the stage for efficient, reliable delivery in subsequent sprints.smooth onboarding for any future teammate, technical debt rules set from day one, operational guardrails that protect momentum, and a mature Retrospective rhythm baked into the team culture. This is how you launch Sprint 1 already optimized.

TEAM VERTEX IN ACTION

Maya leads a Sprint Zero Retrospective with urgency: "We haven't written much code yet, but we've written the system we'll use to deliver all code." Each team member calls out one remove-the-drag improvement:

Jenna – UX: "Wireframes buried in Jira comments slowed dev clarity."

✓ Action: Embed designs directly in stories before Sprint Planning.

Alex – Dev: "Our branch names got inconsistent."

✓ Action: Branch naming conventions enforced through lint checks + PR validation.

Malik – QA: "Testing visibility was low."

✓ Action: A weekly Dev/QA sync meeting is added to stay ahead on edge cases.

Maria – PO: "Design changes at the last second slowed readiness."

✓ Action: "Design Review" becomes a required workflow step before a story becomes 'Ready for Dev'.

Maya – Scrum Master: "Let's make improvement a habit, not a rescue mission."

✓ Action: Retro outcomes tracked as backlog items with owners & completion dates.

They don't wait for later. They upgrade their system now.

AGILE IN MOTION

Team Vertex operationalizes Agile core values:

- **Individuals & Interactions:** Every blocker surfaced early and solved together.
- **Working System Over Documentation:** Their CI/CD + skeleton app is already shipping.
- **Responding to Change:** Retrospective actions implemented immediately, not "next release."
- **Continuous Improvement From Day 1:** Because delay compounds waste.

Velocity tomorrow depends on discipline today.

TEAM REFLECTION

Two truths emerge: Momentum is too expensive to lose and prevention beats correction every time

They leave Sprint Zero with a technical engine running clean, a team aligned in thinking and behavior, and a delivery runway that supports speed. Team Vertex is no longer potential. They're a high-performance delivery unit, activated.

SDLC PHASE 7: MAINTENANCE

Maintenance & Continuous Improvement.
From release to stability.

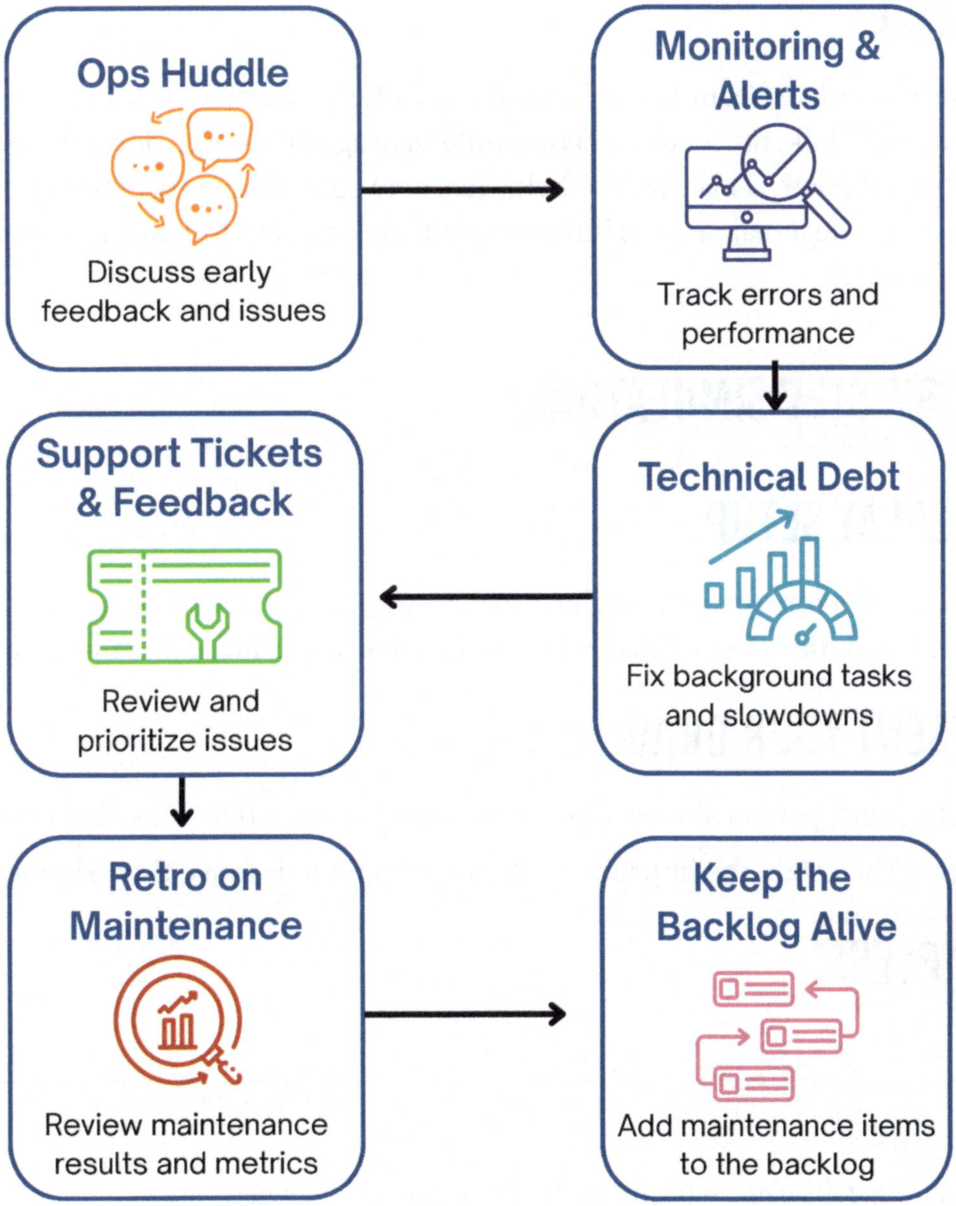

Exercises & Activities

"Retro Like You Mean It" – Hardening the Delivery Engine

Theme: Sprint 0 isn't about features. It's about friction-proofing your system. This retrospective drills into that.

OBJECTIVE:

Participants simulate Team Vertex's Sprint Zero Retrospective. But instead of generic "what went well" talk, the focus is: "What hidden drag will slow us down in Sprint 1 if we don't address it now?" This exercise helps learners spot hidden technical debt before it compounds, create a real process improvement backlog, and link Sprint 0 maintenance to future velocity.

STEP-BY-STEP SIMULATION

1. ROLEPLAY SETUP

Break into 5 roles (or rotate through them): Maya (SM), Malik (QA), Maria (PO), Jenna (UX), and Alex (Dev). Each role gets a Scenario Card (see examples below).

2. PRESENT YOUR DRAG

In character, each person shares one system/process issue that caused friction during Sprint Zero. Then state the improvement you propose to remove that friction.

EXAMPLES:

- Maria (PO):

"Design handoffs came too late, causing stories to sit unrefined."

➤ Proposed Fix: "Add a 'Design Review Complete' checklist item before stories go into Ready."

• Alex (Dev):

"Merge conflicts slowed us down because branch naming wasn't consistent."

➤ Proposed Fix: "Create a branch naming convention and enforce with pre-commit checks."

• Malik (QA):

"No one told me when code was ready to test."

➤ Proposed Fix: "Add a Jira status 'Ready for QA' and auto-notify Slack channel."

3. CAPTURE THE FIXES

Write each fix as: Improvement Title (e.g. "Design Review Step Added"), Owner (e.g. "Maria"), and Completion Target (e.g. "Before Sprint 1 Planning"). Put these into a mock "Retro Outcomes Board" (whiteboard, sticky notes, or Miro template).

4. OPTIONAL BONUS ROUND

Replay the same retro from a future teammate's POV: "How would someone new joining in Sprint 3 struggle because of what we didn't fix now?"

DELIVERABLE

Each group or student outputs a Retro Outcomes Board with 4–5 tangible improvements and a short debrief: "Which of these small issues could have become major Sprint blockers?" "How does this prove Maintenance = Prevention?"

CHAPTER SUMMARY

Sprint Zero ends as it was designed to, with force. The team now has a working skeleton app, a proven pipeline, defined quality gate technical + UX alignment, continuous improvement behaviors, and zero known delivery blockers. Sprint Zero's success metric was never UI. It was confidence and readiness.

Mission accomplished.

WHAT'S NEXT?

The real adventure begins:

Sprint 1 - Planning to Deployment of the first real user value

First stop: **Pet Profile Creation & Login**

Authentication + onboarding = the gateway to all future functionality. Team Vertex is entering the battlefield with: clarity, alignment, a working system, and momentum.

Now... value delivery begins.

SPRINT 1

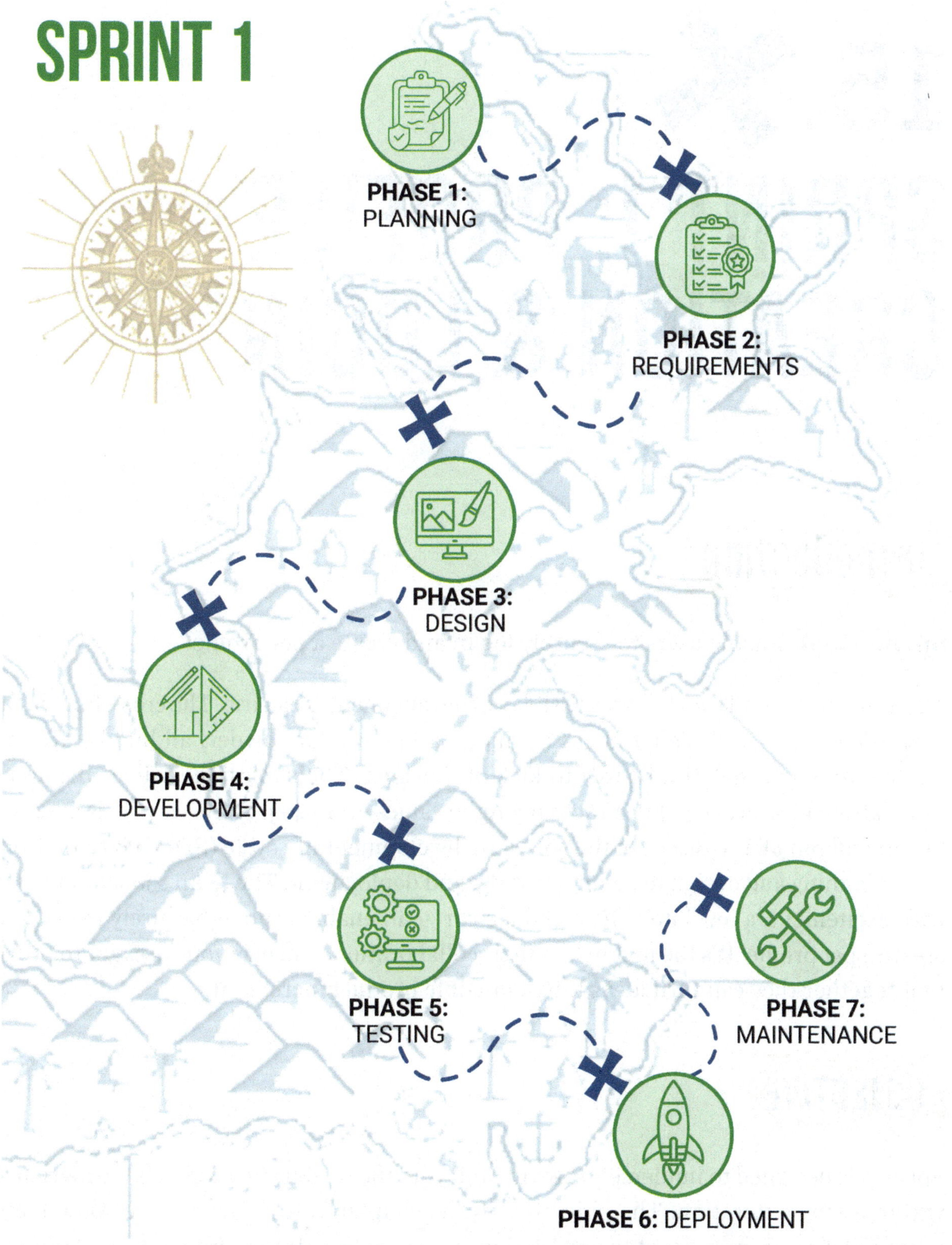
PHASE 1:
PLANNING
PHASE 2:
REQUIREMENTS
PHASE 3:
DESIGN
PHASE 4:
DEVELOPMENT
PHASE 5:
TESTING
PHASE 6: DEPLOYMENT
PHASE 7:
MAINTENANCE

15. SPRINT 1: PROFILE CREATION & LOGIN

INTRODUCTION

Sprint Goal: Enable users to securely log in and create a pet profile.

The atmosphere on Team Vertex's first day of development is electric. The cross-functional team, developers, a designer, a QA engineer, and a product owner, all huddle around a table (and a virtual Jira board) to kick off Sprint 1. Their mission: deliver the app's foundational feature set while practicing Agile values every step of the way. This opening Sprint will put all the pieces of the software development life cycle (SDLC) into motion, from planning and design to coding, testing, and deployment. The team's journey begins with excitement, a few butterflies, and a clear goal: enable users to securely log in and create a pet profile. It's the first test of their collaboration, and everyone is eager to prove that together they can turn ideas into a tangible product increment.

OBJECTIVE

Sprint 1 is designed to immerse the team (and you, the reader) in a full SDLC experience within a single iteration. The objective of this chapter is to show how an Agile team takes a feature from concept to completion in a short time frame, integrating planning, development, testing, and review. By focusing on Profile Creation and Login, the team will demonstrate how Agile practices (like daily stand-ups, iterative development, and retrospectives) align with real-world product development. This Sprint teaches the importance of clear requirements (and the pitfalls of missing them), teamwork under pressure, and continuous improvement. In essence, Sprint 1 offers a crash course in Agile team dynamics and how to successfully deliver a valuable feature while learning from any stumbles along the way.

FEATURE IN FOCUS: PROFILE CREATION AND LOGIN

Profile Creation and Login are the fundamental building blocks of the application. In Sprint 1, Team Vertex sets out to implement these features to enable secure user authentication and pet profile creation. This means a new user should be able to sign up with credentials, log into the app safely, and set up a profile for their pet. It's a critical first step because without user accounts, none of the app's future personalized functionalities can work. The team defines three primary user stories for this Sprint: one for the sign-up process, one for the login process and another for the pet profile setup.

The goal is to ensure that only authorized users can access the app (protecting data) and that each user can create a personalized profile (so, for example, a pet owner can store their pet's name, photo, and other details). By the end of the Sprint, the app should greet users by name (or their pet's name!) and allow them to manage basic profile info. In implementing these, the team must pay attention to user experience (simple, friendly screens) and security best practices like password protection and data validation. Profile Creation and Login may sound straightforward, but as the team will discover, there are important details to get right, and missing any can lead to pesky bugs or security gaps.

TEAM VERTEX IN ACTION

Sprint Planning, Day 1:

The Sprint kicks off with a planning meeting. Product Owner Maria shares the top priority stories from the product backlog.

Maria (PO): "Our first story is 'As a pet owner, I want to create a profile for my pet, so that I can manage their information in the app.' And the second is 'As a user, I want to securely log into the app, so that my pet's data remains private.' These will give us a basic working app to demo."

REQUIREMENTS: BREAKING DOWN THE FEATURE

With Profile Creation and Login set as Sprint 1's focus, Team Vertex translated this feature into concrete requirements.

The team discusses acceptance criteria for each story. They agree that the Profile Creation must include fields like pet name, type, age, etc., and perhaps an option to upload a photo. For Login, they decide users need to provide an email and password to register and then log in, and the system should encrypt or securely handle passwords. They also note some assumptions: for example, using email as the unique user identifier and requiring a minimum password strength.

The Scrum Master, Maya, facilitates the discussion to ensure the stories are understood by everyone and meet the Definition of Ready (DoR). The team asks questions to clarify details:

Malik (QA): "What should happen if a user tries to sign up with an email that's already taken? We should clarify this now."

Maria (PO): "Good catch. We definitely need to prevent duplicate accounts. Let's add an acceptance criterion: Email address must be unique: if an email is already registered, the user gets a friendly error message."

Alex (Dev): "Also, what about password rules? Maybe we say the password must be at least 8 characters and include a number."

Maria (PO): "Agreed. I'll add: Password must be at least 8 chars with a mix of letters and numbers. These details will go into the story."

By asking these questions, Team Vertex is avoiding a planning anti-pattern. They're not assuming anything obvious; instead, they are nailing down specifics now. With acceptance criteria defined, the stories now satisfy the DoR. The team breaks each story into tasks on their Jira board: database setup, UI design, frontend code for forms, backend integration, testing, etc. Each member grabs tasks aligned with their specialty, but they all plan to collaborate closely.

Design & Prototyping:

UX Designer Jenna has already created a quick Figma mockup for the login and profile screens. During the Sprint's first day, she walks the team through the design:

Jenna (UX): "This is the sign-up & login screen. It's simple: our app logo at the top, followed by fields for email and password, and a big, friendly 'Sign Up / Log In' button. And here's the pet profile screen users see after logging in. They can enter their pet's name, upload a cute photo, and fill a couple of fields like breed and birthday."

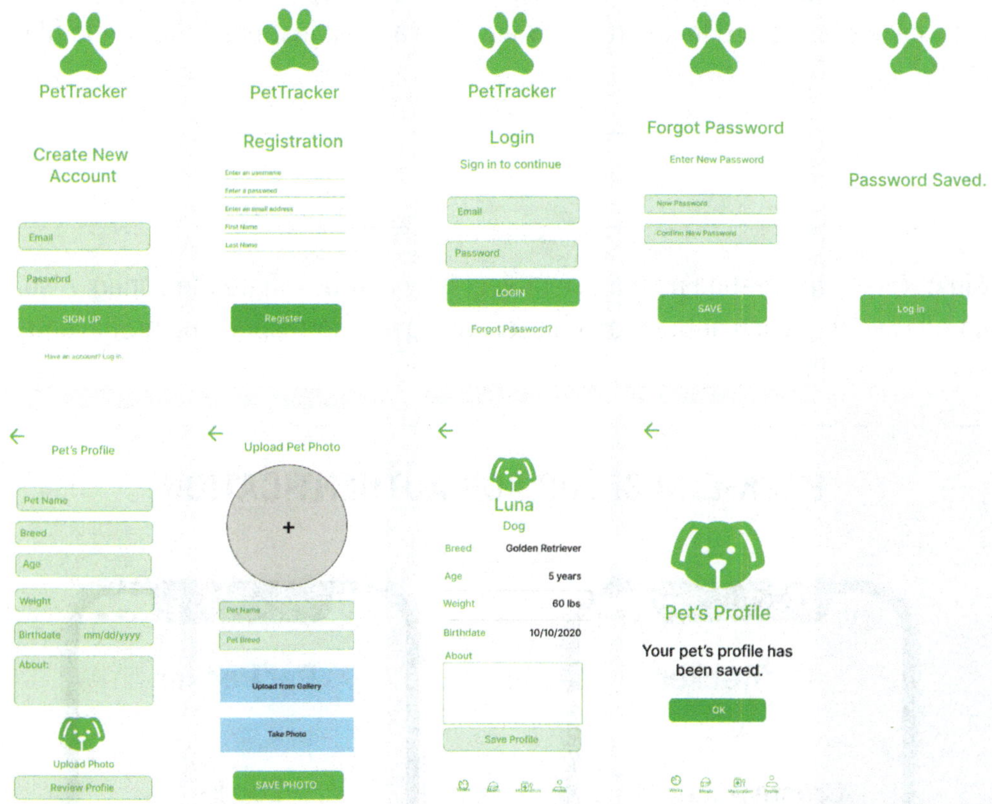

The team gives feedback. You (New Team Member) suggest a minor tweak: "Maybe we should add a confirmation for the password during sign-up to avoid typos." They decide it's a good idea for user experience (even if not explicitly in the story, it aligns with making the app user-friendly and secure). Jenna notes the change in her design. The collaborative design session helps everyone visualize the end product and catch potential UX issues early. By the end of Day 1, the team has a clear plan: tasks are defined, designs are in hand, and everyone knows what to do next.

Development in Full Swing:

Developer Alex takes on the backend work for authentication. Rather than reinvent the wheel, the team opts to integrate Firebase Authentication to manage user sign-up and login.

Alex (Tech Lead): "Using Firebase Auth will save us a ton of time. It handles user accounts securely, and we won't have to store passwords ourselves."

You (New Team Member): "I'll integrate Firebase into our app's front-end. It provides SDKs that make sign-up and login pretty straightforward. I'll just need the config file from Alex."

Alex: "On it. I'm creating a Firebase project now and will share the config. Also, I'll set up rules so that each authenticated user will have a corresponding profile entry in our Firestore database."

BACK-END SETUP FOR AUTHENTICATION

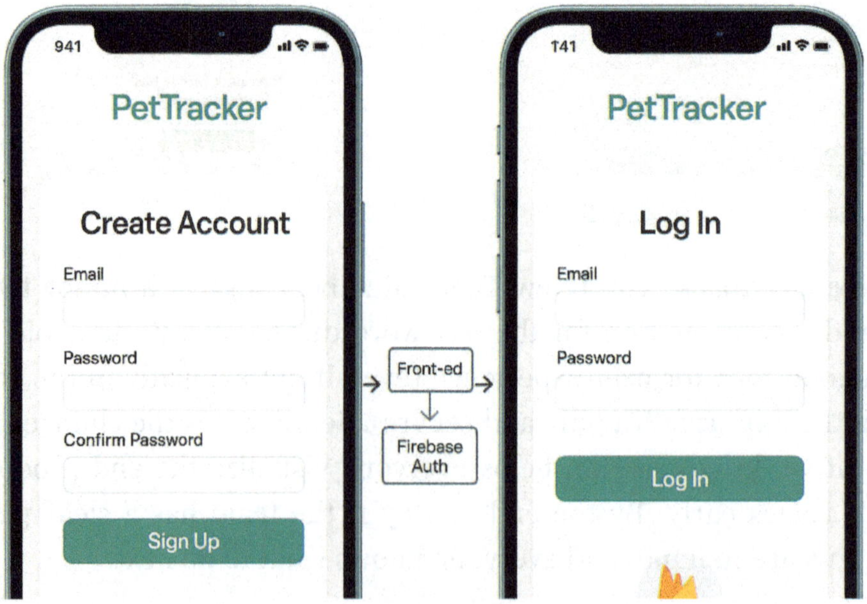

KEY ELEMENTS SHOWN

- **Create Account / Log In screens**: clean teal-and-white UI, ready for front-end integration.
- **Firebase Auth**: highlighting Alex's back-end setup for secure sign-up/login and automatic profile creation.
- **Mini data-flow inset**: depicts front-end → Firebase Auth → Firestore, reinforcing the end-to-end architecture.

XP PRACTICE #1 — CODE REVIEW BEGINS

Before merging their first authentication code, the team practices their new XP rule: Every pull request must be reviewed by at least one teammate before merging. Alex submits the backend code for profile creation, and you are assigned as the reviewer:

You: "I'll review your PR for Firebase setup. I'll check code clarity and make sure the security rules align with what we planned."

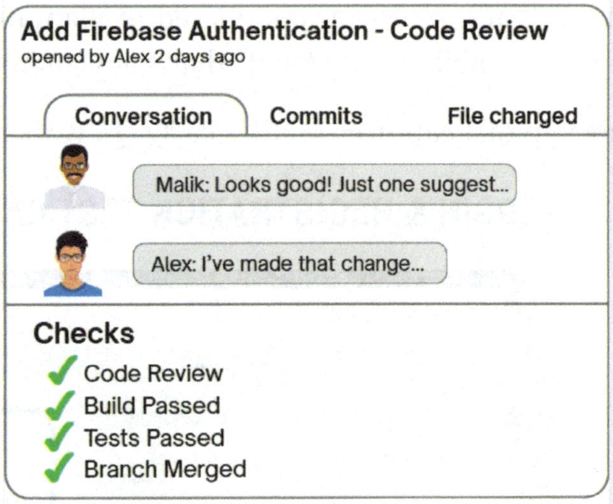

Alex: "Perfect. I'll review yours once your form integration is ready."

Maya reinforces this as a team agreement:

- No direct commits to the main or develop branches
- All work must go through PR reviews
- Changes must meet coding standards and pass automated checks
- Peer review comments are welcomed, not resented

This is the team's first XP Engineering Discipline: Code Review=Shared ownership of clean, secure, and sustainable code. A small moment, but a massive cultural shift: No heroes, hidden mistakes, or soiled knowledge. Just accountability and craftsmanship together.

While Alex and you pair up to get authentication working, Malik (QA) doesn't wait until the end to start his work. He begins writing test cases for the user stories on Day 2.

He drafts scenarios such as:

- "Sign up with a new email should succeed,"
- "Sign up with an already registered email should show an error,"
- "Login with wrong password should fail,"
- "Profile creation with missing info should prompt for required fields."

Malik's proactive test planning not only ensures testing won't be last-minute, but it also sometimes reveals if a scenario was missed in planning. (He's already glad they discussed the duplicate email case up front: that's now in his test list!)

Here's a quick mockup that captures Malik's early QA work alongside the dev effort:

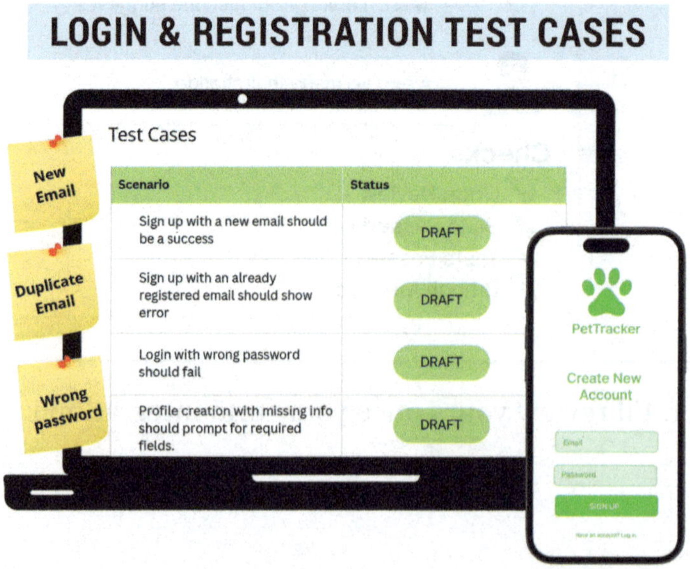

What you're seeing

- **Left (Laptop):** Test-case table in draft status. Exact scenarios Malik outlined, reinforcing "test early" practice.
- **Sticky notes:** Highlight critical edge cases already surfaced (duplicate email, wrong password, etc.).
- **Right (Phone):** Sign-up screen from Sprint 0 UI, visually linking tests to real functionality.

Meanwhile, you build the forms in the app: a simple React (JavaScript) or Flutter (depending on their stack) page for Login/Sign-up and another for Profile creation. You use Jenna's Figma design as a guide, implementing the layout and field validations (like ensuring all required fields are filled). Jenna sits with you to fine-tune the design details; it's a great example of cross-functional pairing: design and development working together to get the feature just right.

By mid-Sprint, the basic functionality is coming together. A user can navigate from the welcome screen to a sign-up form, create an account (you add a nice touch: a little spinner shows while Firebase processes the request), then log in and land on a blank profile page. The profile page UI is ready; now they connect it to the backend. Alex sets up a Firestore collection for profiles. When a user signs up, Alex's code triggers the creation of a profile document with the user's UID (unique ID from Firebase). That way, when the user logs in, the app knows whether a profile exists or not. It's full-stack integration happening iteratively. The team doesn't wait for "backend complete, then frontend," they integrate as they build.

Daily Stand-ups & Communication:

Each morning, the team does a quick stand-up meeting to sync:

Maya (Scrum Master): "Good morning, folks! Let's do our check-in. You, how's the front-end coming along?"

You: "Front-end is on track. Yesterday I integrated the Firebase SDK, and the sign-up form is submitting correctly. Today I'll work on displaying error messages from Firebase. For example, if the user tries an email that already exists, we need to show an error from the backend."

Malik (QA): "I started some exploratory testing on the sign-up flow. I noticed the UI doesn't yet show a confirmation when sign-up succeeds. Maybe we should route the user directly to the profile setup after sign-up? Right now, it's a bit unclear that it worked."

You: "Oh, true. I'll make it so that after successful sign-up, it automatically proceeds to the Create Profile page. Thanks!"

Alex: "On my side, I got the Firestore rules in place last night. Today I'll be connecting the profile form so when a user enters pet info and hits save, it writes to the database."

Maria (PO): "I'd love to see that in action. Ping me when it's ready; I'm curious how the data shows up."

Jenna: "I'm free after lunch to pair with anyone needing UI tweaks. Also, I'll prepare some test images for the profile picture upload to ensure it handles different sizes."

Maya: "Great! And remember, we have only a couple of days before the Sprint Review. Let's keep an eye on the burndown and wrap up any lingering tasks by tomorrow."

Maya notes a minor concern: they're a bit behind on writing unit tests for the backend functions. The team decides that Alex will dedicate a few hours to creating basic unit tests for the sign-up and profile creation logic (for example, a test to ensure duplicate email registration is indeed prevented by Firebase's settings and that profile creation only happens for logged-in users). This is part of their Definition of Done (DoD): all user stories must have passed tests and code review before they are considered "Done."

The Bug Emerges:

Everything seems to be going well until Day 4, when Malik (QA) is doing a full pass of testing on a nearly finished feature. He discovers something worrying: a bug that allows a duplicate email registration to slip through under a certain condition. Specifically, if a user tries to sign up with an email that's already registered but with different letter casing (e.g., using uppercase letters or an extra space), the system fails to recognize it and creates a second account for what is effectively the same email. This could lead to serious confusion or even a security issue (duplicate accounts for one email). Malik immediately brings it up:

Malik (QA) (posting in the team's chat channel with a bug ticket link): "Hey team, I found an issue: I was able to register using TestEmail@example.com even though testemail@example.com was already in the system. It looks like our duplicate email check is case-sensitive, or trimming isn't handled. We might end up with duplicate user profiles."

The message pings everyone. A moment of silence follows as they realize this is a critical oversight, exactly the kind of thing they intended to prevent.

Alex (Tech Lead/Developer): "Oh no, that's on me. I assumed Firebase would handle email uniqueness regardless of case, but it looks like we need to normalize emails or adjust a setting."

You (New Team Member): "We didn't add a front-end check for that either. We should fix it on the backend for sure, and maybe also ensure the front-end lowercases emails before sending."

Maria (PO): "Glad we caught it now. This is definitely a must-fix before we call these stories done. How much time will it take?"

Alex (Developer): "Not too long. Firebase actually treats emails as case-insensitive by default for login, but our logic for creating a profile might have allowed a second profile doc. I'll enforce using toLowerCase() on emails before creation and double-check by attempting a duplicate registration in the Firebase console."

Malik (QA): "I'll re-test as soon as you deploy the fix. Also, let's add an acceptance criterion retroactively: System must treat emails case-insensitively and reject duplicates, so we won't forget this scenario."

Maya (Scrum Master): "This is a good learning moment. It's exactly why we have QA involved early. Let's get the fix in and also discuss in the retrospective how to catch these in planning next time."

The team scrambles a bit, but by the end of Day 4, Alex has implemented a fix: all emails are stored in lowercase and trimmed of whitespace, and Firebase is configured properly to disallow creating accounts with an existing email (now tested thoroughly). You update the front-end to always send lowercase emails as well. Malik verifies the fix; he can no longer create a duplicate account with any email case variations. The bug is resolved.

This little crisis introduced some tension. Alex felt he slipped up, and there's a hint of worry in the team about meeting the Sprint deadline. But by rallying together (and skipping a lunch break or two), they managed to correct course. They also learned a valuable lesson about thorough requirements: nobody explicitly said "emails should be case-insensitively unique" during planning; they all just assumed it. It's a classic oversight that they all vow to avoid in the future by thinking through edge cases.

UAT DEPLOYMENT: FIRST PETTRACKER BUILD

At the end of the Sprint, with development and QA complete, Team Vertex deployed the first increment of PetTracker to the User Acceptance Testing (UAT) environment. This internal "staging" app allowed them to validate the Login & Profile features in a production-like setting before the review.

Scene: The team gathers as Alex prepares the UAT release for the Profile Creation & Login feature.

Maya (Scrum Master): "QA sign-off is done and all tests passed. Alex, are we good to push the profile & login build to UAT?"

Alex (Developer): "Yep! Our CI/CD pipeline is all set. I'm merging the code and triggering the UAT deployment now."

Malik (QA): "I'll run a smoke test on the UAT app as soon as it's live. We should verify that a new user can sign up, log in, and create a pet profile without issues on both iOS and Android."

Jenna (UX Designer): "I'll double-check the UI on a smaller device once it's deployed, just to be sure the screens look good and nothing's off in this

environment."

Maria (Product Owner): "Once the build is up, I'll notify our stakeholder group that they can try out the new feature in UAT ahead of the review meeting."

(A short wait as the automated build and deployment run: the team watches the pipeline status...)

Alex: "Deployment succeeded. The UAT link is live with our latest build!"

Malik: "Great, opening the UAT app now...Sign-up flow works on my device: I created a test account, got to the pet profile screen, entered info, and hit save. All good, the profile was created. No crashes or errors."

Jenna: "UI check on my end looks fine. The logo, form, and confirmation banner are all displaying correctly on the test phone."

Maria: "Wonderful. I'm letting stakeholders know that the app is ready on UAT for them to poke around."

Maya: "Nice job! I'll mark the UAT deployment as complete on our Jira board. We're all set for the Sprint Review. Let's be ready to demo and collect any feedback tomorrow."

By deploying to UAT, the team ensured the increment met the "potentially shippable" standard. It was the first time the fully integrated login system and profile flow ran in an environment similar to production, a big milestone for Team Vertex.

Sprint Review, Day 8:

Come demo time, Team Vertex has a working increment to show stakeholders. In a meeting room (and over video call for remote members), Maria (PO) leads the demonstration of the new features:

1. Users can register an account by providing an email and a password. If they attempt to register with an email that's already in use, a clear error message appears ("Email already registered. Try logging in instead.").
2. Upon successful sign-up, the app automatically logs the user in and takes them to the Create Pet Profile screen.
3. On the profile screen, the user can input pet details (name, breed, etc.) and even upload a photo. The team shows how uploading a cute pet picture instantly updates the profile.
4. After saving, the profile is displayed with the pet's info. The team also demonstrates the login flow: logging out and logging back in with the new account to show persistence.
5. They highlight the security aspect: passwords are not visible and are safely handled by Firebase (no plaintext storage). Maria also notes the validation rules: "Notice we require a number in the password. If I try a weak password, the app rejects it politely."

The stakeholders (including a product manager and maybe a client representative) are pleased. One of them asks, "Can users have multiple pets or just one for now?" Maria explains that for Sprint 1, each user has one pet profile, but the architecture could support multiple in the future, a potential backlog item for another Sprint.

Another stakeholder suggests maybe adding a "Forgot Password" feature eventually; the team nods, acknowledging that it will come in a later Sprint once the basics are in place. Overall, the feedback is positive: the app's first slice of functionality is live, and it works. Sprint 1's goal is achieved: the team delivered a usable feature increment.

As the review wraps up, Scrum Master Maya thanks everyone and sets the stage for the retrospective: "Great job, Team Vertex. Now, let's gather our thoughts on what we can improve. We'll do a Start/Stop/Continue exercise next."

SPRINT 1 UAT DEPLOYMENT: KEY PRACTICES CHECKLIST

Team Vertex treated their first UAT release as more than a handoff—it was a production dress rehearsal.

☑ **CI/CD Pipeline for UAT**

Code merged and deployed through an automated pipeline—no manual build steps.

- ☑ **QA Sign-Off First**

  Only deployed after full QA testing passed. No defects left for stakeholders to catch.

- ☑ **Smoke Test in UAT**

  QA ran a quick end-to-end test to confirm the app was functional in its new environment.

- ☑ **Cross-Device UI Validation**

  UX was reviewed on different screen sizes to catch layout or rendering issues.

- ☑ **Stakeholder Access + Notification**

  The PO informed stakeholders and invited them to test before the Sprint Review.

- ☑ **Prod-Like Environment**

  UAT closely mirrored production for realistic feedback and validation.

- ☑ **Demo-Ready Confidence**

  Deployment to UAT gave the team a stable, working build to showcase in the Sprint Review.

AGILE IN MOTION

Sprint 1 put Agile principles into practice in real time. Here are some key Agile and SDLC concepts that were demonstrated during the Sprint:

Planning Pitfalls and Anti-Patterns: The team nearly fell into a common trap by assuming a requirement was "obvious" (the unique email rule) without explicitly discussing it. This is a planning anti-pattern: never assume, always clarify. The bug with duplicate emails was a direct lesson in the importance of thorough planning and asking "what if" questions. Thanks to a vigilant QA and open team communication, they caught it early. It highlighted why including all team members in backlog refinement (especially QA, who often thinks of edge cases) is vital to avoid gaps.

Definition of Ready & Done Alignment: Team Vertex applied the Definition of Ready (DoR) by ensuring each user story had clear acceptance criteria before development. They realized mid-Sprint that one criterion was missing and promptly updated it, reinforcing that DoR isn't a check-box formality, but a real guide to quality. Their Definition of Done (DoD) also guided them: they knew the story wasn't "Done" until the duplicate email issue was fixed and all tests passed. This alignment with DoR/DoD prevented them from declaring victory prematurely and underscored the value of these definitions in maintaining standards.

Continuous Refinement: Although Sprint 1 was short, the team engaged in quick refinement activities. They refined stories during Sprint planning and even mid-Sprint when new information emerged (e.g., adding the criteria about password rules and email uniqueness). Agile teams refine the backlog continuously. Team Vertex showed this by adjusting scope on the fly (such as deciding to include password confirmation and stronger validation once they thought of it). This adaptability ensured the delivered feature met user needs better by the Sprint's end than it might have if they'd rigidly stuck to the initial (incomplete) plan.

Testing Strategy. "Testing Early and Often": The QA member, Malik, started writing test scenarios from day one. This illustrates the Agile testing mindset: testing is not a phase at the end, but an activity woven throughout the Sprint. By preparing test cases early, he influenced development (Alex and you knew what would be tested and could code accordingly). The team also practiced pair testing when fixing the bug, the developer and QA working side by side to validate the fix. They learned to automate some tests (unit tests for backend logic) to catch regressions early. All these are hallmarks of a sound testing strategy in Agile: involve QA from the start, use tests to drive development quality, and fix issues as soon as they're found.

Sprint Review and Adaptation: The Sprint review was not just a demo, but also a chance to get feedback and adapt the product roadmap. Stakeholder questions about multiple pet profiles and password resets led Maria to note these as future backlog items. This is Agile in motion: after seeing the working product, new ideas and priorities emerged. The team welcomed that input; they didn't treat the plan as static. They'll carry those insights into Sprint 2 planning. By demonstrating a working increment, Team Vertex also built trust with stakeholders, showcasing transparency and the ability to deliver. It's a textbook example of the inspect-and-adapt cycle at the product level.

In summary, Sprint 1 was a microcosm of Agile practice: the team navigated planning pitfalls, upheld their DoR/DoD to maintain quality, refined requirements as they learned more, tested continuously, and embraced feedback at the review. These experiences reinforce why each ceremony and practice exists: they all helped the team deliver a better result and learn how to improve next time.

TEAM REFLECTION RETROSPECTIVE

Sprint 1 concluded with a Sprint Retrospective where Team Vertex reflected on their process and teamwork. Using the "Start / Stop / Continue" format, they identified actionable insights for improvement:

Start:

- Start involving the QA engineer even earlier during story definition and planning. This will help surface edge cases (like the email uniqueness issue) before coding begins.
- Start applying Negative Testing and Edge Case Validation as a standard QA and development practice. Malik's discovery of the duplicate email bug revealed the importance of testing what shouldn't happen, not just the happy path. The team agreed to include these scenarios in both acceptance criteria and test cases."
- Start writing down explicit acceptance criteria for edge cases and error scenarios. The team realized that writing "what should happen when things go wrong" is as important as the happy path.
- Start doing brief mid-Sprint check-ins on requirements. For example, a short backlog refinement mid-Sprint might catch any new understandings or missed details while there's still time to adjust.

Stop:

- Stop assuming "basic" functionality doesn't need discussion. No feature is too obvious to talk through. The duplicate email bug taught them not to assume the simplest parts will take care of themselves.
- Stop last-minute scrambling on test environments. They noted how setting up the Firebase project and test data earlier would have saved time. They'll stop pushing those setup tasks to the eleventh hour.
- Stop skipping unit tests for core logic. There was a moment they almost skipped writing unit tests due to time. They agreed to stop that habit because

those tests could have caught the email case bug sooner.

Continue:

- Continue the great cross-functional collaboration. Pairing between design and development (Jenna and you) and between QA and dev for testing proved extremely helpful. They want to keep that up.
- Continue using Agile ceremonies to stay aligned. Daily stand-ups, in particular, helped surface small issues (like the navigation flow tweak) before they became big issues. The team found these meetings valuable and will maintain the discipline.
- Continue leveraging proven tools and services. Using Firebase Auth was a win; it saved time and provided a secure backbone. The team plans to continue this practice of not re-inventing the wheel for features that can be integrated (as long as they fit the project's needs).

This retrospective conversation was candid and productive. Team Vertex members openly discussed what to improve without blaming, and they made concrete commitments to adapt. By articulating these Start/Stop/Continue points, the team is set to apply the lessons of Sprint 1 to their next Sprint, improving their efficiency and product quality continuously.

Exercises & Activities

"Anyone can talk Agile. Very few can show they can drive a feature through the full SDLC. This Sprint 1 pack does exactly that."

Sprint 1: Profile Creation & Login — Exercise Pack

Portfolio-driven. Practical. Hire-focused.

Part 1 — Workflow / Activity Diagram

Export: 01_ActivityDiagram_ProfileLogin.pdf

Objective

Model the end-to-end behavior for Sprint 1 (Email Sign-Up → Login → Create Pet Profile) before any UI design. This artifact proves you can reason about flow, decisions, and system touchpoints like a pro.

Scope (must include):

- Happy path for Sign-Up, Login, Profile Create
- Error handling (duplicate email, weak password, missing required fields, bad credentials)
- Re-entry loops back to the correct step after errors
- Decision diamonds with clear YES/NO labels
- System callouts for Auth + DB (as notes, not lanes)
- Navigation after success (where the user lands)

Notation Legend (use these shapes)

- Start
- End
- Rounded rectangle = Action/Activity (e.g., "Validate email format")
- Diamond = Decision (Yes/No)
- Small note = System callout (e.g., "Auth: normalize email toLowerCase().trim()")
- Solid arrows = Flow of control

Tools: Figma / FigJam / Miro / Lucid / draw.io / Whimsical

REQUIRED SCENARIOS & BRANCHES

A) Sign-Up (happy path)

- Start → "Enter email + password"
- "Validate format (client)"
- Decision: Valid?

  - No: "Show inline error" → back to input

  - Yes: "Auth: create user (normalize email)"

Decision: Email already in use?

Yes: “Show ‘Email already registered’ → CTA ‘Go to Login’ ” → jump to Login start

No: “DB: create empty profile {uid, createdAt}” → “Navigate to Create Pet Profile”

B) Sign-Up (weak password)

- From “Validate format” → No → inline rule message (“≥ 8 chars + number”) → retry

C) Login (happy path)

- “Enter email + password” → “Auth: verify credentials”

Decision: Auth success?

No: “Show ‘Incorrect email or password’” → back to input

Yes: “DB: fetch profile by uid”

Decision: Profile exists?

No: “Navigate to Create Pet Profile”

Yes: “Navigate to Profile Summary”

Create Pet Profile (happy path)

1. “Enter Pet Name* (required), Type, Birthday, (Photo optional)”
2. “Validate required fields”
3. Decision: Valid?

No: “Inline error under field” → back to input

Yes: “DB: save profile (update)” → “Toast: ‘Profile saved’” → “Profile Summary” → End

E) Optional (stub only)

- “Forgot Password” off Login (mark as Future Sprint)

SYSTEM CALLOUTS (ATTACH AS NOTES)

- Auth: email = toLowerCase().trim(); duplicates are case-insensitive

- DB: profiles/{uid} fields: { petName, breed, birthday, photoUrl?, createdAt, updatedAt }
- Security: Reads/writes only when request.auth.uid == resource.data.uid

BUILD STEPS (15–20 MINUTES)

1. Place Start (●) on the left; End (◎) on the right.
2. Lay out the Sign-Up path across the top row, with decisions for Valid? and Email in use?.
3. Under it, lay out the Login path and merge it into the profile existence decision.
4. On the right, add Create Pet Profile with validation → DB save → Summary.
5. Add error branches that loop back to the right action (not to Start).
6. Add Auth/DB notes where calls occur.
7. Label every diamond with YES/NO on each outgoing arrow.
8. Do a final pass: arrows clean, no crossing spaghetti, minimal text.

QUALITY CHECKLIST (PASS/FAIL)

- Single Start and End node
- All three happy paths (Sign-Up, Login, Profile Create)
- Error handling for duplicate email, weak password, missing required fields, bad credentials
- Re-entry loops go to the correct prior step
- YES/NO labels on every decision
- Auth + DB shown as notes (not lanes)
- Clear navigation after success (Profile Setup / Profile Summary)
- Fits cleanly on one page and readable at 100%

PART 2 — LOW-FIDELITY WIREFRAMES

Export: 02_LowFiWireframes_Annotated.pdf

Objective:

Translate the workflow logic into early UI layouts that show exactly what the user sees at each key moment in Sprint 1.

Your wireframes must make the intent unmistakably clear to:

- Developers building the screens
- QA testing them
- The PO approving them

No guessing. No assumptions.

Screens You Must Include:

Screen	Purpose	Required Elements
Sign-Up / Login	Authen tication entry point	Email + password fields, CTA button(s), error + disabled states
Create Pet Profile	Capture pet details	Pet name (required), breed, birthday, photo upload placeholder, save button
Profile Summary	Confirmation of success	Pet details displayed, CTA for future features (optional stub)

WIREFRAME REQUIREMENTS

Use simple UI shapes like boxes, lines, and circles. Keep it low-fi.

Every screen must include:

- ☑ Required fields marked clearly (ex: * or label text: required)
- ☑ Field validation messages where errors occur (ex: "Password must be at least 8 characters")
- ☑ Disabled button states until required fields are filled
- ☑ Clear navigation outcomes:

- o Success → "Profile Summary"
- o Failure → error messaging + retry on same screen

ADD THESE ANNOTATIONS:

- What triggers each error state
- What backend/system call is linked to each action button
- Where data validation occurs (client side vs. backend)
- Where the user is routed next

Tip: Use short arrows + notes as annotations to avoid clutter.

UX & SECURITY NOTES TO REPRESENT

- Email is treated case-insensitive (normalize text)
- No password revealed ever (mask input)
- Show loading indicator for sign-up/login requests

ACCEPTABLE TOOLS

- Figma
- Miro
- Canva
- Lucid
- Whimsical

- Pen + paper → take a photo (if neat + readable)

READY-TO-TEST DESIGN CHECKLIST

Before exporting, verify:

Check	Yes/No

Every required field marked and validated	
All errors visible exactly where triggered	
Disabled state shown for buttons	
Navigation arrows match workflow diagram	
No decorative styling – purely functional	

If Dev or QA would ask you "Where does this go?" — fix it.

SUBMISSION

Export screens + annotations to one PDF

Name: 03_LowFiWireframes_Annotated.pdf

Upload to Sprint 1 portfolio folder.

WHY THIS MATTERS

This is proof you can design functionally — not artistically — in alignment with:

- User needs
- Business rules
- Technical feasibility
- Testing clarity

This wireframe set becomes the UI contract for the dev + QA team.

Rock-solid wireframes are a Scrum Master's quiet power: They kill confusion before it blossoms into bugs.

PART 3 — USER STORIES + ACCEPTANCE CRITERIA

Deliverable: 3 Completed Stories + Gherkin Acceptance Criteria

Export: 03_Stories_AC_ProfileLogin.pdf

Objective: Write clear and testable User Stories and Acceptance Criteria that describe the behavior for the core Sprint 1 functionality:

- Email Sign-Up
- Login
- Pet Profile Creation

These must be written from the user's point of view and include validation and security controls introduced in Sprint 1.

INSTRUCTIONS

You will create three User Stories:

1. Email Sign-Up
2. Login
3. Create Pet Profile

Each story must include:

- User Story format As a [type of user], I want [goal], so that [benefit].
- Gherkin Acceptance Criteria: Use Given / When / Then format.Minimum 3–4 scenarios per story

Must cover:

- Required field validation
- Password strength rules
- Duplicate email handling
- Successful and unsuccessful flows
- Data persistence after login
- Redirect behavior (post-signup → Create Profile)

No technical language: Write behavior, not code.

Required Content Checklist: Before exporting your PDF, confirm you included:

Requirement	Included?
3 user stories addressing sign-up, login, and profile	
At least 9 total scenarios written in Gherkin	
Error handling scenarios (fields missing, weak password, etc.)	
Duplicate email scenario	
Password masking requirement	
Post-login redirect to profile setup	

PROFESSIONAL TIPS FOR SUCCESS

- Each story must deliver user-visible value
- Each acceptance criterion must be testable
- Think about: What happens when things go right? What happens when things go wrong? and What should the user see every time?

If a tester can't verify it, rewrite it.

SUBMISSION

When complete:

- Export your work as: 01_Stories_AC_ProfileLogin.pdf
- Upload to your portfolio folder
- Be prepared to walk through ONE scenario in your Sprint Review role-play

PART 4 - TEST CASE MATRIX

Export: 04_TestCases_ProfileLogin.csv or 04_TestCases_ProfileLogin.pdf
(Your choice — spreadsheet or formatted document)

OBJECTIVE

Demonstrate that you can design a full testing scope for a user-facing feature before developers ever write code. This builds your QA partnership muscle as a Scrum Master:

- You anticipate risk
- You protect the Definition of Done
- You ensure user experience is verified, not assumed

REQUIREMENT: MINIMUM 8 TEST CASES

Cover both happy path and edge cases

Suggested coverage groups:

- Valid sign-up with new email
- Duplicate email sign-up blocked (case-insensitive)
- Weak password validation enforced
- Missing required profile fields blocked
- Login with correct credentials
- Login with wrong password isn't allowed
- Successful profile creation saves data
- Photo upload error (optional stretch case)

Tip: Good portfolios include at least one error path per story

REQUIRED COLUMNS

Column	Description
TC ID	Unique identifier (ex: AUTH-01)
Preconditions	What must be true before testing
Steps	Exact actions a tester performs
Expected Results	What success looks like
Pass/Fail	Leave blank – used during execution

Write steps so a new tester could follow without confusion.

TESTING STANDARD GUIDANCE

Use Gherkin logic to shape Expected Results even if not explicitly formatted that way:

Example: When a user enters an email already registered then the app should display "Email already registered." If you cannot define expected behavior...

your team has not clarified the requirement yet. That's an Agile red flag so, document it and raise it in refinement.

FORMATTING TIPS

- Keep steps short and precise
- Use numbering for readability
- Expected results must be binary (either met or not)
- Avoid design notes — focus on behavior

SUBMISSION CHECKLIST

Check	Yes/No
Covers sign-up + login + profile creation	
Includes happy and error paths	
Exact messaging written for errors	
Tests protect acceptance criteria	
Spreadsheet format is clean and scannable	

WHY THIS MATTERS

Test cases showcase your mastery of quality, not just process.

Strong Scrum Masters know how to verify outcomes, prevent ambiguity from entering development, shield teams from rework and chaos later

This artifact proves you can lead with clarity from Day 1.

PART 5 — UAT DEPLOYMENT WALKTHROUGH

Export: 05_UAT_Walkthrough.pdf

OBJECTIVE

Simulate what a User Acceptance Testing validation looks like in the real world regardless of whether you have a functional app yet. Stakeholders don't test code. They test capability, flow, and confidence. This exercise proves you can guide that experience.

WHAT YOU'RE DELIVERING

A step-by-step walkthrough of the Login + Profile Creation journey using: prototype screenshots, annotated success checks, stakeholder-style comments, and a clear PASS on the increment. Think like you're presenting during a Sprint Review.

SCREENS TO INCLUDE (MINIMUM 4)

Screen	Proof Required
Sign-Up	New user creation visible (fields completed)
Login	Successful authentication navigates onward
Create Pet Profile	Required fields + validation visible
Profile Summary View	Saved info appears correctly

If you don't have UI yet:

Use annotated Figma mockups or hand-drawn wireframes with photos. Still counts.

REQUIRED ANNOTATIONS

Each screen must show: Short action summary (e.g., "User enters valid email and password"), expected vs. actual behavior, green check if passed, user experience notes (stakeholder voice!), and navigation routing shown (arrows or captions)

Example Stakeholder Notes:

- "UI is clean and intuitive — allow future upload cropping?"
- "Consider password visibility toggle?"
- "Form spacing looks mobile-friendly — good!"

END-TO-END VERIFICATION

Include 1 short validation section at the end:

Question	Question
Does account persist after logout/login?	Profile loads
Any blockers preventing flow completion?	None
Ready for stakeholders to use in UAT?	Yes

This reinforces Definition of Done.

PRESENTATION STYLE TIPS

- **Use blue checkmarks for pass states**
- **Make each step feel like a story unfolding**
- **Avoid technical jargon — speak in business value language**
- **Clean layout — centered images, short captions**

Your role is the narrator of confidence.

SUBMISSION CHECKLIST

Requirement	Done?
Screens sequenced logically	
Each action includes Expected vs Actual outcome	

At least one stakeholder comment per screen	
End-of-flow validation included	
PDF export is polished and presentation-ready	

WHY THIS ARTIFACT MATTERS

This shows employers you can:

- Lead UAT with professionalism
- Ensure increments are release-ready
- Bridge the gap between technical and business worlds

Stakeholders buy trust. This document proves your team earns it.

PART 6 — STAKEHOLDER SPRINT REVIEW SCRIPT

Export: 06_SprintReview_Script.md

Format: Clean, executive-ready bullet script

OBJECTIVE

Show that you can lead a Sprint Review with clarity, confidence, and business alignment — not a technical monologue.

Stakeholders need three things: What was delivered,, why it matters, and what's coming next.

This script proves you can guide that narrative.

DELIVERABLE REQUIREMENTS

Write a 1–2 minute demo script as if you are the Scrum Master or Product Owner presenting:

- Authentication + Profile creation
- Working navigation from sign-up profile
- Confirmation that data persists

Use language stakeholders understand like value, safety, speed, trust, and user experience

STRUCTURE + PROMPTS

1. Opening (Context & Sprint Goal)

"We set out to enable users to securely log into the app and create a personalized pet profile."

2. What We Delivered (Facts)

- New user sign-up with secure password handling
- Login path that returns users to their saved profile
- Profile creation with pet name, breed, and photo upload
- Validation and error states to protect from mistakes

3. Why It Matters (Business Value)

Explain the capability unlocked: "Now users can safely access their pet's data from any device — the foundation for all future app features."

4. Demo Highlights (Show, Don't Tell)

Guide audience visually using screens or screenshots — 3–5 bullet actions

5. Feedback & Adaptation (Proof of Agility)

State one improvement driven by stakeholder or QA input (e.g., duplicate email fix, password rule clarity)

6. Risks + Next Steps (Transparency)

- Identify one manageable risk + mitigation plan
- Appetite for future enhancements (Forgot Password, multiple pets)

7. Close Confidently (Ready for More)

Confirm that the increment meets the Definition of Done and is ready for stakeholders to explore in UAT.

PROFESSIONAL DELIVERY TIPS

- Keep it tight — 120 seconds max

- Speak benefits before features
- Never apologize for what isn't built yet
- Use future framing:

"This unlocks _______ in Sprint 2."

Your tone should be **confident, crisp**, and **user-centered.**

SUBMISSION CHECKLIST

Requirement	Done?
1–2 minute script length	☑
Covers delivery + value + feedback + risk	☑
Uses non-technical language	☑
Clear narrative that aligns to Sprint Goal	☑
Markdown format	☑

WHY THIS ARTIFACT MATTERS

Hiring managers want communication strength, stakeholder leadership, and confidence on the mic. This script is proof you can run the room and the board.

Part 7 — Portfolio README One-Pager

Export: 07_Portfolio_README.md

Format: Clean Markdown for LinkedIn / GitHub / Job Applications

PART 7 — PORTFOLIO README ONE-PAGER

Export: 07_Portfolio_README.md

Format: Clean Markdown for LinkedIn / GitHub / Job Applications

OBJECTIVE

Summarize this Sprint's deliverables into a single, professional narrative that sells your role, skills, and impact. This becomes a portfolio artifact that hiring managers can skim in 20 seconds and instantly understand what you can do.

1. PROJECT + SPRINT OVERVIEW

"PetTracker – Sprint 1: Login + Profile Creation Feature."

Include sprint goal (1 sentence) and what was delivered (short bullets).

Example phrasing: "Delivered secure user authentication and personalized pet profile creation to support future app features."

2. MY ROLE + ACTION TAKEN

Make it clear and confident:

- What tasks did YOU directly contribute to?
- How did you support collaboration and flow?
- Any leadership or initiative moments?

Example: *"Designed workflow logic, created user acceptance criteria, supported Dev + QA collaboration by clarifying edge cases early."*

3. KEY SKILLS DEMONSTRATED

Select 3–5 relevant competencies such as:

- Agile Requirements / User Story Writing
- UX Collaboration
- Test Case Design + Acceptance Testing
- Stakeholder Delivery + Sprint Review
- Workflow Visualization
- Release Readiness

Keep bullet list short + powerful.

4. EVIDENCE OF WORK

Insert screenshots or attach links to:

- Workflow diagram ✅
- Acceptance criteria ✅

- Wireframes ☑
- Test matrix ☑
- UAT walkthrough ☑

Label each with caption + your name. This is where "I contributed" becomes proof.

5. LESSONS LEARNED + ADAPTATIONS

Brief and reflective.

Examples: *"Documenting error paths earlier prevents rework."*

"QA involvement from Day 1 is a game-changer."

SUBMISSION CHECKLIST

Requirement	Done?
Exactly 1 page	☑
Includes screenshots or links	☑
Uses clear business language	☑
Shows YOUR impact (not "the team"	☑
Exports cleanly to PDF if needed	☑

WHY THIS ARTIFACT MATTERS

This one page becomes:

- A portfolio hero piece
- A hiring conversation starter
- A tangible story demonstrating how you deliver value in a real Agile team

It makes you unforgettable.

FINAL FOLDER STRUCTURE

Sprint1_ProfileLogin_Portfolio/

- ☑ – 01_Stories_AC_ProfileLogin.pdf
- ☑ – 02_ActivityDiagram_ProfileLogin.pdf
- ☑ – 03_LowFiWireframes_Annotated.pdf
- ☑ – 04_TestCases_ProfileLogin.csv
- ☑ – 05_UAT_Walkthrough.pdf
- ☑ – 06_SprintReview_Script.md
- ☑ – 07_Portfolio_README.md

Exercises & Activities

EXERCISE 1: REFINEMENT ROLE PLAY — "SHOULD THIS STORY BE READY?"

SCENARIO:

Team Vertex is reviewing a new user story during Backlog Refinement: "As a pet parent, I want to receive vaccine reminder notifications so that I never miss important vaccination dates." Sounds promising.but as the conversation unfolds, tension bubbles up. Not everyone agrees this story is Ready for Sprint Planning. You'll step into each team member's shoes to determine if it's ready or if it needs more refinement first.

OBJECTIVE

Practice identifying what makes a user story truly 'Sprint Ready' by analyzing gaps in clarity, dependencies, and cross-functional input during backlog refinement.

INSTRUCTIONS

Read the Dialogue Below

Review what each team member says during refinement. Pay attention to clues about missing information, dependencies, or risks.

Decide as a Team

Based on what you've read, decide:

- Should this story move into the next Sprint?
- Or should it stay in refinement (or trigger a Spike)?

List the Actions Needed for Readiness

If you decide it's not ready, identify exactly what needs to happen to make it Ready.

REFINEMENT ROLE PLAY

Maria (PO): "I understand the business urgency, but we need complete acceptance criteria and feedback from Design and QA before this can move forward."

"I'll work with stakeholders to clarify the reminder frequency and delivery method (email, push, etc.)."

Alex (Dev): "There's a technical dependency we need to explore—can we define that before we estimate accurately?"

"I'd recommend a Spike to research how reminders sync with the existing notification system."

Malik (QA): "I can't confirm test coverage without defined acceptance criteria or edge cases. What scenarios should we account for (e.g., expired vaccines, duplicate reminders)?"

"Let's add validation rules like required vaccine name and reminder date."

Jenna (Designer): "I haven't reviewed the UX flow. Do we expect reminders to appear in the dashboard or as a notification? I'll need to mock that up."

"Let's schedule a quick design sync to finalize placement and interaction."

Maya (Scrum Master): "Let's not rush it into the sprint. This story lacks cross-functional readiness and carries risk."

"We can timebox a Spike to resolve the dependency and add it to the next refinement session."

YOUR TASK

1. Ready or Not?

Should this story move into the Sprint backlog?

- Yes — It's Ready for Sprint Planning
- No — Keep it in Refinement or Add a Spike

Explain your reasoning in one or two sentences:

2. Refinement Actions:

List the next steps your team should take before this story becomes Sprint Ready.

EXERCISE 2: ESTIMATING BY EFFORT — GETTING A FEEL FOR STORY POINTS

SCENARIO:

Team Vertex is new to Agile estimation, and it's time for their first Planning Poker session. The Scrum Master (Maya) explains that story points don't measure time — they measure effort and complexity.

The team will estimate three simple PetTracker stories by comparing them to one another.

OBJECTIVE

Learn to size stories relatively based on effort, complexity, and risk, not on how long they'll take.

INSTRUCTIONS

1. **Start by Choosing a Baseline**

   Pick the simplest story and call it a "3-point story."

   All other stories will be estimated relative to that one.

2. **Review the Three Stories**

   Read the short user stories below and discuss which one feels simple, moderate, or more complex.

3. **Decide the Point Range**

   Choose between 1, 3, 5, or 8 points based on how much effort or uncertainty the story represents.

4. **Explain Your Reasoning**

   Write one sentence describing why you gave that estimate.

STORIES TO ESTIMATE

STORY A: ADD A NEW PET

> *"As a pet parent, I want to add my pet's name, species, and birthdate so I can start using the app."*

Your estimate (1–8 points): ________

Why?

STORY B: LOG A FEEDING

> *"As a pet parent, I want to record what my pet ate and when, so I can track*

feeding times."

Your estimate (1–8 points): ________

Why?

STORY C: VIEW MY PET'S DAILY SUMMARY

> *"As a pet parent, I want to see all of today's feedings and walks in one dashboard, so I can review activity at a glance."*

Your estimate (1–8 points): ________

Why?

CHAPTER SUMMARY

In Sprint 1, Team Vertex successfully delivered the first crucial features of their project, secure user login and pet profile creation, effectively implementing a full SDLC slice from planning to UAT deployment. Along the way, they navigated challenges that tested their Agile practices and team collaboration. They started with clear goals, and by the end of the Sprint, they had a working product increment in the hands of stakeholders.

The team's accomplishments this Sprint included:

- Setting up a robust authentication system (leveraging Firebase) that handles user registration and login securely.
- Creating a user-friendly profile page for pet owners to input and manage their pet's information.
- Ensuring the app meets basic quality criteria (validation rules, error handling, persistence of data) for these features.

More importantly, the lessons learned were invaluable. The duplicate email bug incident taught them the importance of clarifying requirements and thinking through edge cases early. It highlighted how essential open communication and cross-functional input are in Agile teams; a QA perspective during planning can save a lot of pain later on. Team Vertex also got to see the benefits of their Agile ceremonies: daily stand-ups kept them coordinated, and the Sprint review provided feedback that will shape future work. By holding a retrospective and committing to improvements, they set the stage for continuous growth as a team.

By the end of Sprint 1, Team Vertex not only built a piece of their app but also grew stronger as a unit. They have a shared success under their belt, a clearer understanding of their process, and confidence that they can tackle whatever comes next. The pet app now has its foundational login and profile feature. A small step for the product, but a giant leap for a team in its first Sprint together.

WHAT'S NEXT?

With the first Sprint done and dusted, Team Vertex turns its eyes to the horizon of Sprint 2. They've established user accounts and profiles; the app's groundwork is in place. But what new challenge will the next iteration bring? Perhaps the team will build on this foundation, maybe enabling new interactive features for pet owners or tackling an even more complex user story. One thing is certain: they will carry forward the lessons from Sprint 1 to work smarter and smoother.

Sprint 2 promises to introduce a fresh feature and, with it, new opportunities for learning and improvement. Will the team be able to sustain their momentum and up their game? What new Agile hurdles or technical puzzles might they encounter as they expand the app's capabilities? Stay tuned for the next chapter, where Team Vertex dives into their next Sprint: armed with experience, facing new goals, and ready to continue their Agile adventure. The story of Team Vertex continues in Sprint 2: what do you think will happen next?

SPRINT 2

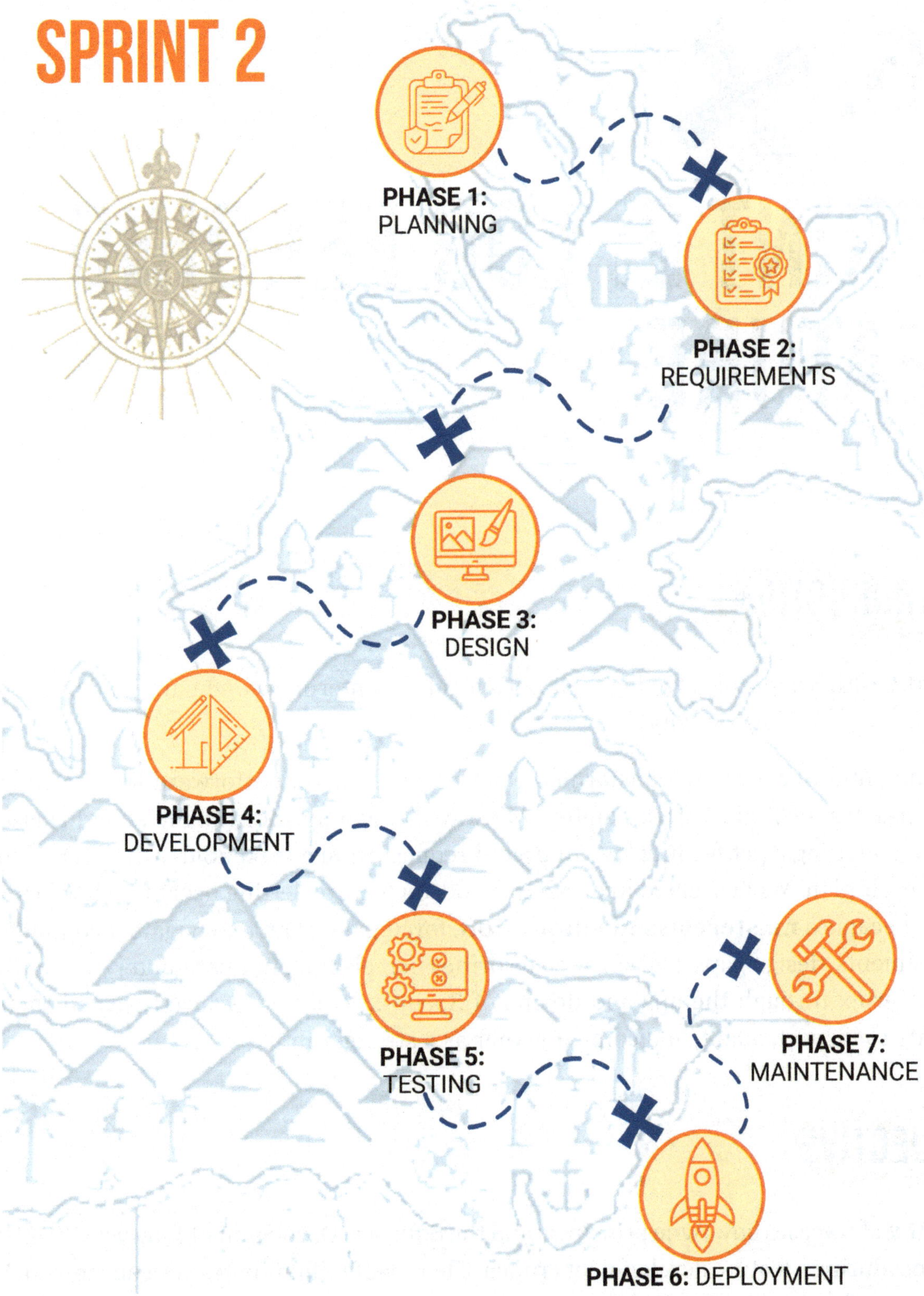
PHASE 1:
PLANNING
PHASE 2:
REQUIREMENTS
PHASE 3:
DESIGN
PHASE 4:
DEVELOPMENT
PHASE 5:
TESTING
PHASE 6: DEPLOYMENT
PHASE 7:
MAINTENANCE

16. SPRINT 2: LOG FEEDING FEATURE

INTRODUCTION

Sprint Goal: Enable users to log their pet's meals by entering a food name and portion size (the core Feeding Activity).

A sudden mid-Sprint twist transformed what began as a straightforward feature build into a true test of Team Vertex's agility. Midway through Sprint 2, as the team focused on implementing a pet feeding log, an urgent request from a stakeholder threatened to derail their plan. What seemed like a simple add-on feature, letting users log their pet's meals, became a **masterclass in adaptation**, forcing the team to navigate incomplete requirements, design gaps, and surprise priorities in real-time. In this chapter, we follow Team Vertex through the ups and downs of Sprint 2, where pressure revealed their capacity to inspect, adapt, and come together as a stronger unit.

OBJECTIVE

Sprint 2 showcases how Agile principles and the Software Development Lifecycle (SDLC) intersect during a feature development crunch. This chapter illuminates key concepts such as **story refinement and slicing**, maintaining focus on the **Sprint Goal** amid change, cross-functional collaboration during **design handoff**, the importance of thorough **QA testing beyond the happy path**, and **continuous improvement** through retrospectives. By the end, you'll see how the team integrated planning, development, and testing practices to deliver value under pressure, illustrating how a Scrum team "*does* Agile" and also "*is* Agile" embracing change, communication, and learning at every step.

FEATURE IN FOCUS: FEEDING ACTIVITY

The **Feeding Activity** (or *Feeding Log*) feature allows pet owners to record their pet's meals by entering what the pet ate and how much. On the surface, it's a simple form with two fields, but it's one of the app's most essential functions, forming the core of daily usage. By enabling meal logging, this feature helps users track their pet's diet and feeding schedule over time. Its significance is high: not only does it encourage frequent user engagement (logging each meal), it also lays the groundwork for future enhancements like nutritional analysis or quick-add meal presets. Team Vertex understood that delivering a *reliable* feeding log was critical for user trust and app value, so getting it right was a top priority. Sprint 2's goal was clear, as Product Owner Maria put it during planning.

> "BY THE END OF THIS SPRINT, A USER SHOULD BE ABLE TO INPUT THE FOOD NAME AND PORTION SIZE. THAT'S OUR CORE OUTCOME."

TEAM VERTEX IN ACTION

Planning & Refinement: Slicing the Story, Nailing the Details

After successfully deploying login and profile setup in Sprint 1 to UAT, Team Vertex kicked off Sprint 2 refinement with a (JAD) session led by Maria. The team gathered to discuss exactly how "log pet meals" would work in practice. Maria reiterated the Sprint Goal, recording a pet's meal with food name and portion, and opened the floor for questions. Very quickly, the team realized *clarity* would be key, as even a simple feeding log raised many questions:

Validation rules: "What validations would be needed to prevent blank or nonsensical entries?" Jenna (UX) asked, noting the form should guide proper input.

Numeric input details: "Should portion size accept decimals or whole numbers only?"

Alex (Developer) wondered, concerned about how to store portion data.

Data integration: "How would the data integrate with the pet profile backend?" Malik (QA) chimed in, ensuring the new data would align with existing profile info.

User feedback: "What message would confirm a successful save?" Maya (Scrum Master) prompted, reminding them that user confidence in saving data matters.

Through this discussion, Team Vertex reached a consensus on the **minimum scope** for the Feeding Log:

- A simple form with **two fields**: Food Name and Portion Size.
- **Inline error handling** for missing or invalid inputs (no pop-ups, errors should appear near the fields).
- An **API connection** to store the meal log in the backend database.
- A brief **confirmation message** (a "Meal saved!" toast) to reassure users that the log was successful.

To visualize the user's experience, the team even sketched a basic **workflow diagram** mapping out the feeding log process from start to finish. In this flow: the user enters the food name and portion size, the app validates the inputs, and if all is well, it submits to the backend and displays a confirmation. Anticipating error paths (like invalid input leading to inline error messages) in this diagram helped the team uncover edge cases early. With a shared vision of the workflow and agreement on scope, Sprint 2 began with confidence and clarity.

LOG FEED WORKFLOW DIAGRAM – HAPPY PATH

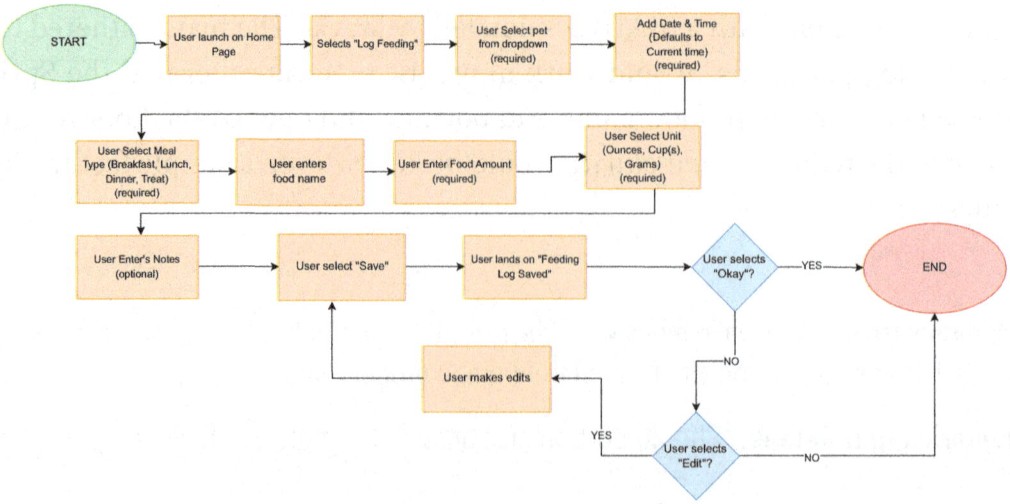

REQUIREMENTS: BREAKING DOWN THE FEATURE

As the team moved into **backlog refinement**, they uncovered more complexity. What started as one user story began to **expand into several** once they examined all the required behavior. Maria and Malik took the lead in breaking down the "Log Feeding" epic into discrete, testable user stories:

- **Story 1: Basic Entry**. *"As a pet parent, I want to enter the food name and portion size so I can track what my pet eats."* This covered the core form inputs.
- **Story 2: Error Handling**. *"As a user, I want to receive feedback if I leave a field blank or enter invalid data."* This addressed validation and error messages.
- **Story 3: Confirmation Feedback**. *"As a user, I want to see a confirmation that my food log was saved so I feel confident my data is recorded."* Covering the success message and form reset.
- **Story 4: Presets (Future)**. "As a user, I want to select from a list of common food presets to save time." (Marked as not in MVP scope, to be handled in a later Sprint).

LOG FEED FEATURE TO STORIES BREAKDOWN

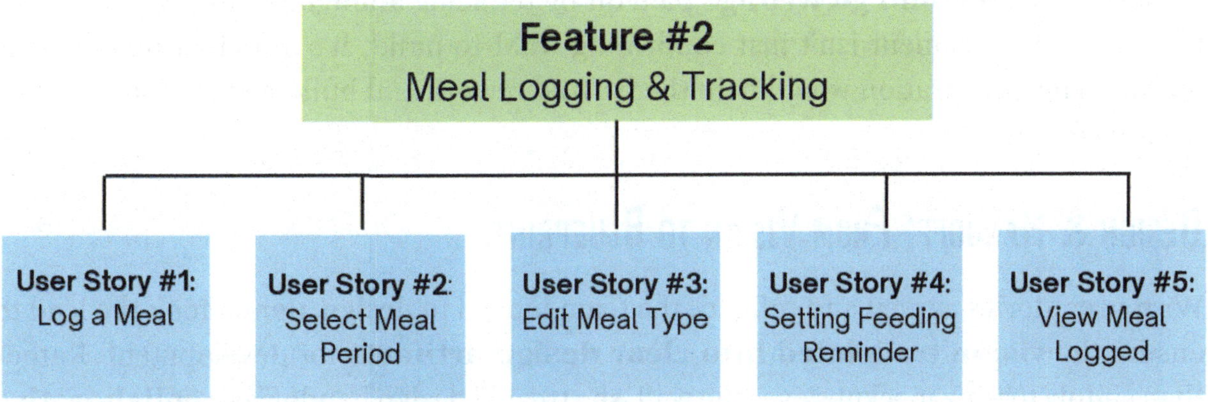

This **story slicing** made estimating the effort easier and allowed parallel progress on parts of the feature. During refinement, Malik raised a crucial question that the initial user story hadn't explicitly answered: **"What happens if a user leaves the portion field blank? Or types in emojis instead of a food name? Are we validating for that?"** This "happy path" user story had glossed over input validation and error states, details that could make or break the user experience. His question prompted the team to step back and enumerate **Acceptance Criteria (AC)** for the feeding log. Together, they defined clear rules to close the gaps:

- The user **must enter both** a food name and a portion size. No blank submissions.
- Portion size must be a **positive number** (no negative numbers, letters, or emojis).
- If either field is left blank or has invalid input, an **inline error message** should explain the issue.
- Upon a valid submission, the app should **display a success message** and **clear the form** for the next entry.
- Any invalid input must **block the Save action** and highlight the problematic field so the user can fix it.

Jenna immediately went into the prototype to annotate these behaviors on the design screens, so that the acceptance criteria would be visually represented. Alex updated the Jira tickets for each story to include these ACs, ensuring everyone from developers to testers had a single source of truth on expected behavior. By the end of refinement, what looked like a "small" feature was now thoroughly specified. The effort spent up front to clarify **"what could go wrong"** paid off by reducing ambiguity later. Team Vertex learned that refinement isn't just confirming what to build, it's anticipating potential pitfalls. This preparation would help them avoid rework and build a more robust, user-friendly feature.

Design & Handoff: From Vision to Blueprints

With user stories agreed and refined, the team moved to design. Jenna took the lead to ensure the **vision translated into clear design artifacts** for development. Rather than simply tossing mockups over the wall, she treated design handoff as a **collaborative bridge** from planning into action.

First, Jenna updated the **Figma** prototypes for the Feeding Log screen. In the shared design workspace, she included detailed annotations so that nothing was left to guesswork:

Notes on **input behaviors**: for instance, how validation errors should appear inline below the fields, and the exact text constraints (e.g., maximum character count for the Food Name field).

Definitions of **button states**: how the "Save Meal" button should look and behave when disabled, during loading (upon tap), and after a successful save (perhaps briefly showing a checkmark or disabled state).

Any **style or layout specifics** that developers might need: ensuring consistency with the app's design system established in Sprint 1.

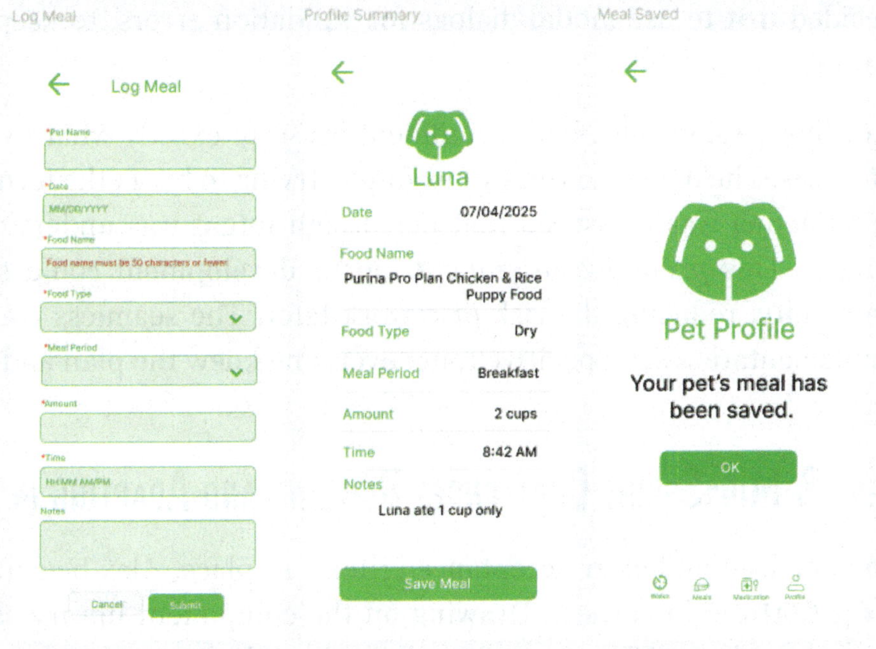

Each Jira story was then tagged "**Design Complete**" once these assets were ready. Importantly, Jenna linked the relevant Figma artboards directly in the Jira tickets, so Alex (Dev) and Malik (QA) could easily find the exact design for each user story. This traceability is a small step that prevents miscommunication. Everyone always refers to the same design source.

To solidify the shared understanding, Jenna scheduled a short **handoff walkthrough** meeting with Alex and Malik right after Sprint Planning. In this 20-minute sync, she didn't just review the visuals; she highlighted the functional expectations tied to those designs. Key discussion points included:

Field logic: Jenna clarified that *"Portion Size can include decimal values (for half-cup measurements, etc.) but no alphabetic characters or symbols."* Alex took note to enforce numeric input and filter out any other characters.

Validation behavior: They confirmed the Save button should remain **disabled until both fields are valid**, preventing any chance of sending incomplete data.

Success confirmation: Instead of a disruptive pop-up, the design called for a subtle **toast message** at the bottom, saying "Meal saved!", and an automatic form reset after submission. Malik asked if the toast should disappear on its own; Jenna confirmed it would fade after a couple of seconds, per standard UX practices.

Error handling approach: All errors would be **field-specific and inline**, directly under the field in question (e.g., "Please enter a portion size" under that input). They explicitly decided **not** to use modal dialogs for validation errors, to keep the context clear to the user.

By the end of this session, Alex had a confident grasp of exactly what to build, Malik knew what test cases he'd need to run (for example, trying to break the form by entering bad data), and Jenna was reassured that her design intent was understood by those implementing it. This *proactive alignment* meant development could start without lingering ambiguity, reducing the risk of rework later. The seamless transition from design to implementation set a positive tone: everyone knew the plan and their part in it[29].

Development & Mid-Sprint Challenge: Building and Adapting in Real Time

With designs finalized and clear acceptance criteria in place, Alex began **coding the Feeding Log feature** in earnest. Drawing on the component library established in Sprint 1, he reused existing UI components where possible for consistency (for example, the text field styles from the profile form). The form comprised two input fields: **Food Name** (text) and **Portion Size** (numeric), and a **"Save Meal" button**. Following the requirements:

- The Save button was initially **disabled** and would only enable once both Food Name and Portion inputs were non-empty and valid.

- Alex set up **real-time validation** on each field: typing into the fields would live-check the input (e.g., portion must be a number) and show an inline error immediately if something was off. This way, by the time a user hits "Save", the form is already validated.

On the backend side, he integrated with the existing API endpoint POST /log-meal. The app would send the entered data (foodName, portionSize) along with an identifier for which pet profile this meal belongs to (retrieved from the user's session).

LOG FEED – AMOUNT FIELD BUG

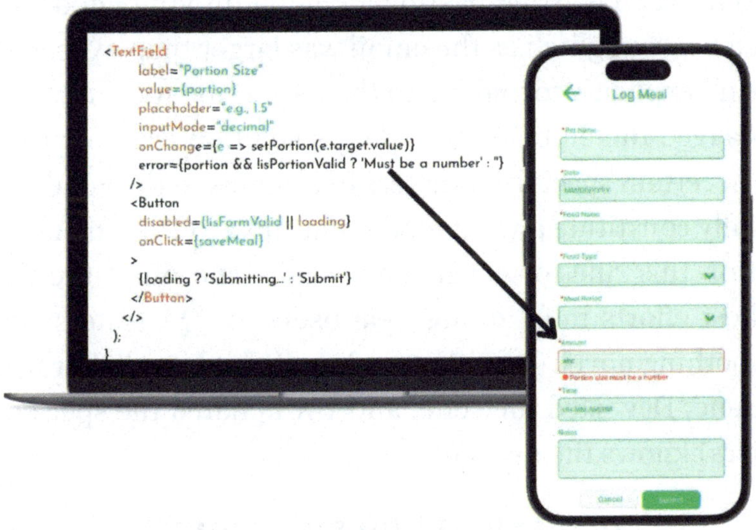

Early in development, this integration hit a snag: the server responded with errors. In the next Daily Scrum, Alex raised this as a blocker: *"The API is rejecting our meal log submissions. I suspect the payload structure might be wrong."* This quick transparency paid off; Maria realized the **API schema** expected slightly different field names and perhaps an authentication token that wasn't being sent. The team discussed it right then, and Maria provided the correct expected JSON format. Alex adjusted the code accordingly, and the API calls succeeded on the next try. This incident underscored why **daily stand-ups are crucial:** a potentially days-long impediment (chasing an API bug) was resolved in minutes by surfacing it early and getting the right eyes on the problem.

As development continued, Alex built out handling for various **edge cases** beyond the basic user story:

- Prevent any submission if inputs are empty or invalid (the button enabling logic covered this, plus an extra check server-side just in case).

- Reject non-numeric entries in the Portion field. E.g., if someone tried to type "two" or put an emoji, that input would be ignored or cause a validation error.

- Once "Save" is tapped, immediately **disable the button** (and perhaps show a spinner) to prevent double-submission while waiting for the server response.

- If the API call fails (due to no internet, server error, etc.), display an **error message** so the user isn't left wondering. (For example, "Could not save right now, please check your connection.")

Midway through the Sprint, just as the team settled into a rhythm, **Malik (QA) discovered a quirk:** when he tried pasting an emoji into the Food Name field on his test device, it caused a layout glitch as the emoji was larger than expected and broke the input styling. It wasn't explicitly forbidden in the requirements, but clearly an edge case to address. Malik flagged this immediately. Alex's solution was to implement a simple **input filter** to block certain special characters like emojis in the Food Name field (since an emoji doesn't really constitute a valid food name anyway). Jenna updated the design guidelines to mention this rule for consistency. This kind of on-the-fly adjustment is common, despite best efforts in planning, real users (or QA testers simulating them) will always find something novel. Team Vertex handled it by collaborating across roles: QA identified the issue, Dev fixed the code, and UX updated the spec so everyone (and future team members) knows the decision.

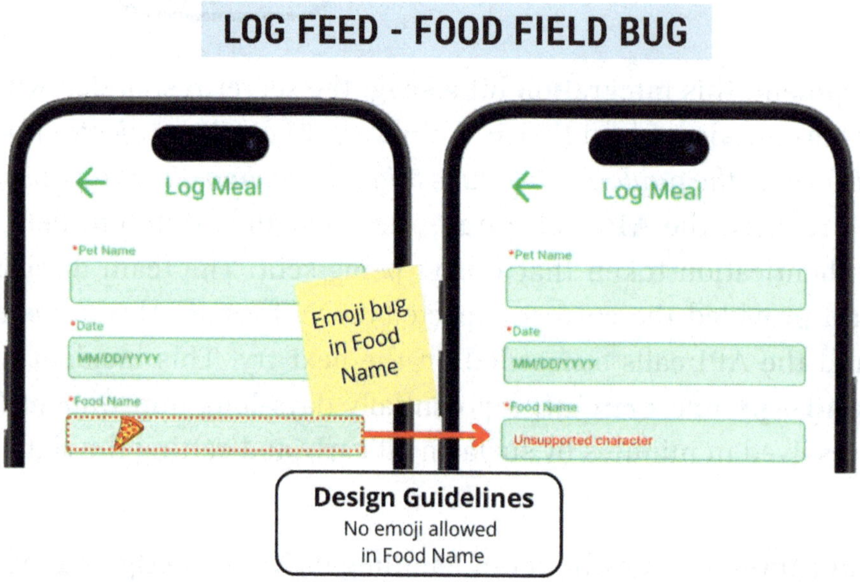

Just as the team resolved these refinement issues and regained momentum, Sprint 2 threw another curveball: a **stakeholder's surprising request**.

XP Practice #2 – Pair Programming in Action

As the Feeding Log form evolved, so did the team's approach to collaboration. Rather than working in silos and tossing issues over the wall, they leaned into their second XP discipline: "Two brains, one keyboard" – Pair Programming.

Alex and Malik began pairing actively on form validation and backend integration, especially after the emoji bug surfaced.

- Malik tested edge inputs in real time

- Alex adjusted logic and added input filters
- They discussed user experience trade-offs together
- Jenna joined for a "UI polish" round, shaping decisions live

This wasn't formal or forced. It was responsive, efficient, and focused.

Pairing allowed the team to reduce handoff waste, catch issues faster, make immediate decisions across roles, and build trust through real-time feedback.They didn't need a process doc, they just needed a second set of eyes, a shared goal, and a willingness to build side-by-side.

On Day 7 of the Sprint, Maria (PO) received an urgent message from the company's Head of Marketing. There was an upcoming pet industry trade show, and Marketing was eager to unveil a new feature: **"Pet Profile Sharing."** This feature would allow users to generate a shareable profile of their pet: essentially, a cute social media card. The stakeholder was pushing to get at least a basic version of Pet Profile Sharing done *immediately*, even though the team was in the middle of Sprint 2, focused on the Feeding Log.

Maria knew this request put the team in a tough spot. That afternoon, she called an impromptu meeting with the entire Scrum Team and Scrum Master, Maya. The dilemma was clear: **add a new high-priority item mid-Sprint** and potentially compromise the Sprint, or **defer the request** and possibly disappoint a stakeholder. The team discussed their options candidly, essentially mirroring the choices Scrum allows in such scenarios:

- *Option A:* **Accept** *the new Pet Profile Sharing story into Sprint 2* *immediately, reallocate capacity, and drop some originally planned work to make room.*

- *Option B:* **Politely decline** *the change mid-Sprint. Explain the need to stay focused on the Sprint Goal (feeding log) and* **commit to scheduling** *Pet Profile Sharing as a top item in the next Sprint.*
- *Option C:* **Cancel Sprint 2** *altogether and re-plan it entirely around the Pet Profile feature as the new goal.*

Maya reminded everyone of the guiding principle from Scrum: a mid-Sprint change should only be accepted if it doesn't jeopardize the Sprint Goal. Otherwise, it's safer to defer or, in extreme cases, replan the Sprint.

After thoughtful discussion, Team Vertex unanimously leaned toward **Option B**: *stay the course*. The Feeding Activity was a committed goal and nearly built; dropping it could mean ending the Sprint with nothing fully done. The stakeholder's request, while important, wasn't so urgent that it trumped the current goal. They reasoned it would be better to finish the feeding log now and start Pet Profile Sharing in Sprint 3 with full attention.

Maria, with Maya's support, communicated this decision back to the stakeholder promptly.

Maria (PO): "We understand the Pet Profile feature is valuable. However, our team is in the middle of Sprint 2, focused on delivering the feeding log. Introducing a new item now would risk not delivering either properly. We will gladly prioritize Pet Profile in Sprint 3, aiming to have it ready as soon as we realistically can."

The stakeholder, initially anxious for a quick yes, accepted this reasoning once they realized forcing everything at once might lead to poor quality or missed deadlines.

By handling the request this way, Team Vertex demonstrated two core Scrum values: **Commitment** (to the Sprint Goal) and **Openness** (being transparent with the stakeholder about what was possible). Maria immediately added the Pet Profile story to the **Product Backlog** with high priority for the next Sprint, to ensure it was not forgotten. The team also reminded themselves (and the stakeholder) that truly critical needs can be handled, e.g., by swapping out work or canceling a Sprint, but those should be rare exceptions, not the norm. In this case, deferring the request protected the team's focus and ultimately both features' quality.

Outcome of Development: Despite the mid-Sprint turbulence, by the end of Sprint 2's development phase, Team Vertex had achieved what they set out to do:

- The Feeding Log **form UI was fully implemented** and responsive to different devices.
- All the **validation rules** from the acceptance criteria were working: the app would not accept blank or nonsensical inputs.
- The **backend integration** was successful: meal logs were indeed being saved via the API, with proper handling of both success and error responses.
- The **Save button's interactive states** (disabled/enabled, etc.) matched what Jenna had designed, providing a slick user experience.

And importantly, the team had navigated a potential crisis (the scope change request) without derailing the Sprint.

Throughout development, daily stand-ups and open communication proved vital. Small blockers like the API schema mismatch were surfaced and solved quickly. The mid-Sprint request was tackled as a team decision rather than a unilateral call, reinforcing trust. By the time coding was complete, Team Vertex was ready to verify that their work held up under scrutiny.

Testing & QA: Catching the Edge Cases

Once the feeding log feature was code-complete, it was deployed to a **shared QA environment** for testing. Malik, as the QA lead, took on the task of validating the feature across different conditions and devices. While the core functionality (entering two fields and saving) worked on the "happy path," Malik specifically set out to **break the feature** in ways an end user might, focusing on edge cases that developers might not have fully anticipated. His testing checklist included:

Trying to **submit the form with blank fields** (should be prevented, with errors shown).

Entering **invalid characters** (like emojis, symbols, or letters) in the **Food Name or Portion** fields to see if any slipped past validation.

Typing an obviously wrong input (like text in the Portion field, which should be numeric) to confirm that the app rejects it.

Tapping "Save" repeatedly in quick succession to simulate an impatient user: would the app accidentally allow duplicate submissions?

Verifying that after a successful submission, the **fields reset** properly, and the success message disappears appropriately.

Malik's thorough testing uncovered two critical **bugs** that the development logic hadn't accounted for:

1. **Empty Submission Allowed:** Through an unusual sequence of actions, something like rotating the phone's screen and quickly focusing in/out of fields, Malik found that the Save button, which was supposed to stay disabled for blank inputs, *became enabled erroneously*. This allowed him to hit "Save" with no data entered, and the app attempted to send an empty log (clearly not intended).
2. **Duplicate Logs on Rapid Taps:** If the Save button was tapped multiple times very quickly, multiple requests could indeed be sent before the first response came back. This resulted in **duplicate meal entries** being recorded for one user action.

LOG FEED – BUG REPORT TICKETS

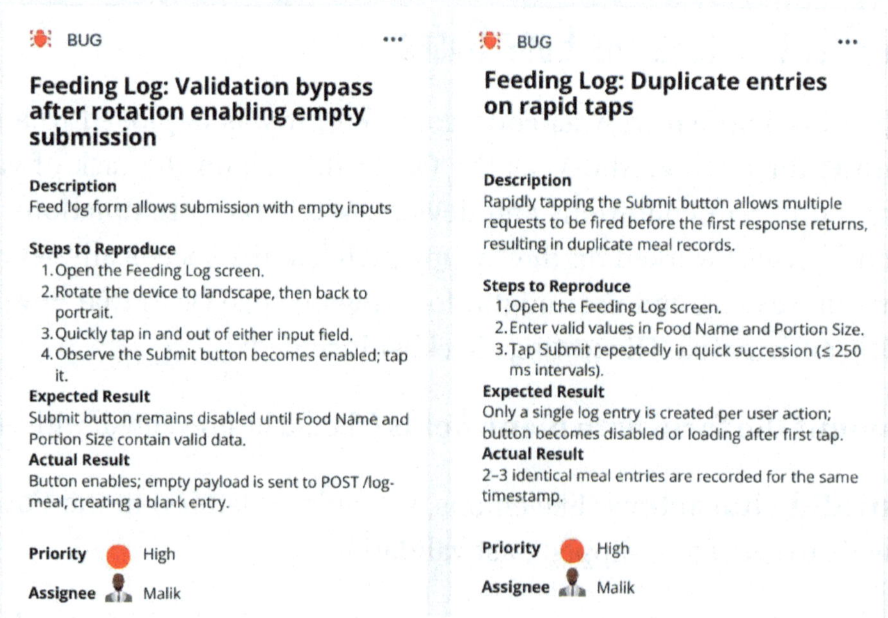

These weren't scenarios the average user might encounter often, but they were possible, *and* could undermine trust in the app (imagine thinking you logged one meal and seeing two appear!). Alex jumped on fixing these issues as soon as Malik reported them. To resolve the **duplicate submissions**, he implemented an **input "debounce"** and a client-side lock: essentially, once Save is tapped, further taps are ignored until a response returns. This ensured one server call at a time. For the **empty submission glitch**, Alex adjusted how the Save button's enabled/disabled state was managed at the UI level, including re-checking validity on certain UI events (like screen rotate) so it could never turn on unless data was present.

After these fixes, Malik **retested on both Android and iOS simulators**, deliberately trying the same tricks that uncovered the bugs. In both cases, the issues were resolved: no more ghost submissions or duplicates.

With these bugs resolved and confidence restored, Malik proposed a new improvement to the team's QA process: establishing a basic regression test suite to guard against similar issues re-emerging in future sprints. The team identified key components—such as the Save button behavior, form validation logic, and API response handling—as shared elements that could impact other features. By creating a short checklist of these core behaviors, they ensured that any fix or new feature would be tested against them automatically.

To reduce manual rework and increase confidence before each UAT handoff, Malik also began laying the groundwork for test automation. Using a mobile test framework compatible with the PetTracker stack, he drafted a smoke test that mimicked the happy path flow for the Feeding Log: entering valid data, submitting the form, and verifying success messages. This script became the team's first automated check—a seed that would grow into a broader safety net in upcoming sprints. With the major bugs squashed, Malik went through a final test case checklist to make sure all acceptance criteria and typical use cases passed:

- **Required field validation**: confirmed you can't save without both fields filled.
- **Input sanitation**: verified that disallowed characters (like % or emoji in Food Name, or letters in Portion) were either prevented or caused the proper inline error.
- **API response handling**: simulated a server error to see an error message, and a successful save to see the success confirmation.
- **Button state transitions**: ensured the Save button accurately reflected

validation state (disabled vs. enabled) and the loading state on submit.

- **User experience on success/failure**: checked that the success toast appears and disappears as intended, and that on a failure, the user is informed without the app crashing or freezing.
- **Regression coverage**: verified that recent fixes did not impact other form components or previously stable features (e.g., Login screen, Settings form).
- **Automated smoke test**: implemented and executed a script to validate the happy path submission flow in both Android and iOS environments.

This rigorous testing phase reinforced an invaluable lesson: **testing beyond the happy path is crucial**. The issues Malik found were not failures of core logic, the main code worked, but rather *timing and interaction* issues that only emerge when you test like a real user (or an overly zealous one!).

Thanks to QA's work, the feeding log feature became much more resilient before any end user ever saw it. By the end of Sprint 2's **Testing** phase, all defined acceptance criteria were met and verified in the QA environment, and the team had high confidence moving into the UAT deployment.

UAT DEPLOYMENT: FEEDING FEATURE ON STAGE

After a whirlwind of development and thorough testing, Sprint 2 reached a point where the Feeding Activity feature was ready to showcase. The team deployed the fully tested Feeding Log to the UAT environment to ensure everything worked in a production-like setting before meeting with stakeholders.

Scene: Team Vertex gathers on the afternoon before the Sprint Review to push the new feature to UAT.

- **Maya (Scrum Master):** "All test cases passed, and the Feeding Log is looking solid. Are we ready to deploy to UAT for the demo build?"
- **Alex (Developer):** "Ready when you are. Kicking off the CI/CD pipeline for the UAT release now."
- **Malik (QA):** "Great, I'll do a quick smoke test on the UAT app once it's live. I want to double-check that logging a meal works end-to-end in that

environment."

- **Jenna (Designer):** "I'll verify the UI in UAT too, especially that the error messages and the 'Meal saved!' toast display properly on a real build."
- **Maria (Product Owner):** "Once you two give a thumbs up, I'll notify the stakeholders that they can try the feature in UAT. They might appreciate getting a sneak peek before the review."

(Minutes later, the automated build succeeds and the UAT app is updated.)

- **Alex:** "Deployment complete. The UAT build has the Feeding feature now. Here's the link for our testers."
- **Malik:** "I've logged a test meal on UAT and it went through perfectly. Got the confirmation message and saw the entry appear in the database logs. Also intentionally left the portion blank once; the app correctly blocked the save and showed the error. Looks good!"
- **Jenna:** "UI and toast look good on my phone. No layout issues either. We're all set."
- **Maya:** "Excellent. UAT deployment is done and dusted. I'll update Jira and our board. Now let's get ready to wow the stakeholders with this demo in the Sprint Review."

Having the Feeding Log on UAT gave the team extra confidence. It ensured that by the time they were presenting, the feature had already been exercised in an environment similar to production. They could head into the review knowing the demo would reflect a realistic user experience.

Sprint Review: Demonstrating the Increment

Sprint 2 culminated in a **Sprint Review** meeting where Team Vertex showcased the new Feeding Activity feature to stakeholders and collected feedback. After deploying the fully tested feature to the UAT (User Acceptance Testing) environment (using their CI/CD pipeline for a smooth release), the team had a live build ready for demo. For many on the business side, this was the first time seeing a working end-to-end user story in the PetTracker app beyond the login screen, marking a significant milestone.

During the review, Maria led a **live demonstration** of the Feeding Log on a projector screen or via screen share. She walked through the primary use case:

- Selecting a pet (if needed) and navigating to the "Log Meal" screen.
- Entering a sample **Food Name** ("Dry Kibble") and **Portion Size** ("1.5 cups"), intentionally demonstrating both a valid entry and what happens if you try an invalid one (she showed that leaving a field blank would immediately show a red-bordered field with an error message like "Food name required," illustrating the validation).
- Clicking the **Save** button and showing the resulting behavior: the button greys out momentarily (to indicate processing), then a subtle **confirmation message** "Meal saved!" appears.

She pointed out how the form cleared itself, ready for another entry, proving that the data was accepted.

The stakeholders could see the **validations in action**. For example, Maria demonstrated that if she tried to type a letter into the Portion field, nothing happened (it refused non-numeric input), and if she erased the portion, the Save button was disabled again.

This not only proved the feature worked, but also reinforced to everyone one of Agile's core principles:

"WORKING SOFTWARE IS THE PRIMARY MEASURE OF PROGRESS".

Instead of just talking about what they did, the team was able to show a tangible, functioning piece of software that delivered value.

Following the demo, the session opened up for stakeholder use and feedback. Some business sponsors and even potential beta users at the meeting got access to the UAT build to try it on their own devices. As they played with the feeding log, feedback **emerged organically**. One sponsor remarked, *"This is great! Could we add preset options like '¼ cup' or '1 cup' so common portions can be logged faster?"* This was a new idea: essentially a convenience feature to save frequent measurements (very much in line with the "Presets" story Team Vertex had identified as a future enhancement). Maria thanked them for the suggestion and immediately noted it on the **product backlog refinement board** for consideration in an upcoming Sprint. The team clarified that while it wasn't in the current release, they saw its value for usability and would evaluate it soon. This real-time feedback loop is exactly what Sprint Reviews are for: the product increment spurs new ideas and improvements, ensuring the product roadmap stays aligned with user expectations.

By the end of the review, the **Feeding Activity feature was accepted** by the Product Owner (Maria) as done, and stakeholders were satisfied that it met the Sprint's goal. They were pleased to see no critical issues in the demo and that the team had even handled an unexpected mid-Sprint request professionally. The deployment checklist was ticked off with pride:

- ☐ The UAT build was accessible to testers and stakeholders via a shared link.
- ☐ The feeding log functionality had been **verified by the Product Owner** and demoed live without problems.
- ☐ Stakeholders had reviewed it, and their feedback was captured for the next iterations.
- ☐ No show-stopping bugs appeared post-deployment, indicating a quality increment.

This phase validated not only the product increment but also the team's ability to **deliver incremental value** and adapt based on real-world input. Sprint 2 ended with a sense of accomplishment: the team had turned a rough concept into a working feature that users could actually benefit from. But equally important was what came next: reflecting on how the Sprint went and how to improve.

AGILE IN MOTION

Sprint 2 put several Agile and SDLC principles into practice, highlighting how a Scrum team navigates the complexities of real-world development:

- **Story Slicing & Incremental Delivery:** Team Vertex avoided a monolithic approach by splitting the Feeding feature into multiple smaller **user stories**. This allowed them to deliver a **vertical slice** of value (basic logging, error handling, confirmation) in one Sprint and defer nice-to-haves (like presets) to later. Story slicing made estimation more accurate and enabled parallel work, illustrating the Agile idea of breaking down work into the smallest valuable increments.
- **Refinement and Definition of Ready:** The team's thorough backlog refinement, where they collaboratively defined **acceptance criteria** for each scenario, is a textbook example of achieving a solid **Definition of Ready** before

development. By fleshing out validation rules and edge cases in advance, they ensured the team knew when a story was truly ready to pull into a Sprint. This prevented mid-implementation surprises and scoped the work to what "Done" really meant. Agile teams use refinement to build shared understanding, and Sprint 2 shows why: the initial user story was *too vague*, but refinement made expectations explicit.

- **Continuous Communication and Collaboration:** Throughout development, the team leveraged Scrum ceremonies to stay aligned. **Daily Scrums** were used effectively to surface and remove blockers (e.g., Alex raising the API issue and resolving it the same day with Maria's input). There was also tight collaboration outside the typical ceremonies: Jenna's **design walkthrough** with Alex and Malik ensured cross-functional alignment from the get-go, embodying the Agile principle of *individuals and interactions* over rigid processes. QA wasn't siloed until the end; Malik was involved in refinement and design discussions, and Alex and Jenna responded quickly to his findings mid-Sprint, a great example of the **whole-team approach** to quality.
- **Responding to Change vs. Following a Plan:** The mid-Sprint stakeholder request was a classic Agile trial. Team Vertex showed that while they are **responsive to change**, they also uphold the importance of the **Sprint Goal**. Scrum's guidance is that a team can adjust scope mid-Sprint if (and only if) it doesn't sabotage the Sprint Goal. By choosing to defer the **Pet Profile Sharing** feature to the next Sprint, the team preserved focus on delivering the value they'd promised this Sprint, demonstrating agility *with discipline*. They also kept the stakeholder engaged by transparently explaining the trade-offs, turning a potential conflict into a learning moment. This scenario underlines how Agile isn't about blindly saying "yes" to all change; it's about negotiating scope smartly and maintaining a sustainable pace.
- **Definition of Done & Quality Focus:** Team Vertex adhered to a robust **Definition of Done (DoD)** before calling the Feeding feature complete. We see this in how they handled testing and acceptance: the feature wasn't considered "done" until it passed all acceptance criteria, had no critical bugs, and was deployable to UAT. They also implicitly followed DoD items like code being merged, peer reviewed, and integrated into the CI/CD pipeline for deployment. This ensured the increment truly met a potentially shippable standard by Sprint's end. Moreover, their emphasis on testing beyond the happy path aligns with an Agile quality mindset: testing is integrated into the Sprint, not a phase after it. Malik's findings and Alex's quick fixes within the same Sprint exemplify **fast**

feedback loops to build quality in, rather than inspecting it after.

- **Continuous Improvement:** Perhaps the most important Agile principle on display was **Inspect and Adapt**. The team didn't wait for a big phase-end review to improve; they were adapting *within* the Sprint (refining acceptance criteria when a gap was found, adjusting design when an issue arose, negotiating scope when priorities shifted). And at the end, during the **Sprint Retrospective**, they identified concrete improvements to implement immediately (standardizing validation rules, creating a reusable error handling pattern). This willingness to tweak their way of working is the hallmark of an Agile team. They treat processes and practices not as fixed, but as continually evolving based on lessons learned.

Through Sprint 2, we see Agile values and principles not as abstract ideals, but as *practical tools*. From maintaining **focus** on a Sprint goal, to fostering **openness** and **courage** in raising issues (like Malik questioning requirements or Alex admitting a bug), Team Vertex applied Scrum values daily. The integration of SDLC stages, planning, requirements, design, development/coding, testing, deployment, maintenance/reflection, into the timeboxed Sprint demonstrates how Agile frameworks like Scrum complement the classic development cycle. By doing a little of *each* step every Sprint (instead of big upfront phases), they delivered a working feature and learned from it sooner. Sprint 2 is a clear illustration of the Agile mindset: be flexible, be collaborative, and never stop improving the product or the process.

TEAM REFLECTION

After the dust settled, Team Vertex convened for a candid **Sprint Retrospective**. The mood was a mix of pride and realism: they had met the Sprint goal and navigated turbulence, but there were lessons to capture. In this retrospective conversation, the team identified a couple of **recurring pain points** that had popped up not just in this Sprint but even earlier:

Inconsistent validation logic across features: Jenna and Malik noted that the rules and methods for input validation seemed to be reinvented for each new form (login vs. profile vs. feeding) and sometimes implemented slightly differently by different developers. This inconsistency could lead to user confusion and more bugs.

Duplicate effort in error handling: Alex realized he had coded inline error message functionality from scratch for the feeding form, even though something similar was done for the login form. There wasn't a shared pattern, so each form's errors behaved a bit differently. This was time-consuming and risked inconsistency.

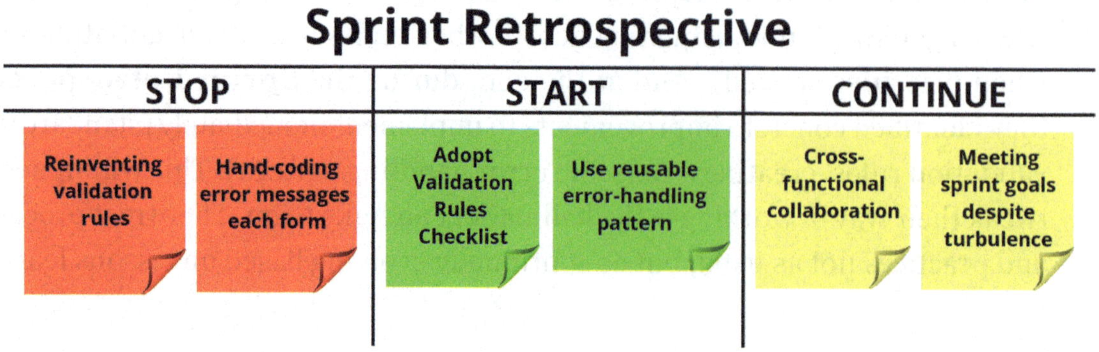

Recognizing these issues, the team agreed on **actionable improvements** to carry forward:

1. They created a **Validation Rules Checklist** (spearheaded by Malik and Jenna), which was essentially a standard list of validation requirements to consider for any new form. It included items like:

- [x] "All required fields must be marked and cannot be blank,"
- [x] "Enforce max character limits where applicable,"
- [x] "Use numeric-only for numeric fields,"
- [x] "Keep the Save/Submit button disabled until inputs are valid."

This checklist would now serve as a *Definition of Done* item for design and development: before a story involving a form could be marked complete, it had to meet all these standard validation checks. By systematizing this knowledge, they would reduce oversights and ensure a consistent user experience across the app. Alex took the initiative to document a **reusable error-handling code pattern**. He wrote a snippet (in the project wiki or Confluence page) demonstrating how to implement inline field error messages in a standardized way. This included how to structure the error objects, how to trigger them on validation failures, and even a bit of styling guidance to match the design.

The team agreed to adopt this pattern going forward, effectively creating a library of solutions for common problems. The next time they build a form, developers can copy or import this error-handling approach rather than writing it anew each time. Consistency would improve, and development might even speed up thanks to these ready-made patterns.

Reflecting on the team's journey in Sprint 2, you can sense that Team Vertex is **maturing**. They moved past just "forming" and into "**storming**" and "**norming**" phases of team development. Early on, unexpected challenges (incomplete requirements, mid-Sprint change requests) created a bit of *storm*: moments of friction or uncertainty that tested the team's unity. However, how they responded is telling: rather than descending into blame or chaos, they pulled together to solve problems, which is a sign of emerging trust. By the end of the Sprint, when they sat down to improve their process, they showed strong **norming** behavior, establishing new norms (like the validation checklist and coding standards) that will guide their work in the future. This indicates a shift toward higher performance; the team is not just focusing on the product anymore, but also on how they work together.

Maya, the Scrum Master, highlighted this in the retrospective: *"Notice how we handled conflict and change this Sprint; we addressed issues openly and came out stronger."* The team agreed. Sprint 2 had moments of tension (e.g., telling a stakeholder "not now" is never easy), but they managed those tensions constructively and without personal conflict. By explicitly improving their process (toolkit enhancements) and aligning on best practices, Team Vertex turned Sprint 2's lessons into lasting team knowledge.

In the Tuckman model, this is the hallmark of **Norming**. The team is setting its norms, resolving tensions quickly, and enhancing cohesion. They're not a brand-new team feeling their way anymore; they're developing a rhythm and a growing confidence in each other.

Exercises & Activities

SPRINT 2 — EXERCISE PACK

Feature: Log Feeding

Focus: Validation + Data quality + Continuous improvement

Part 1 — Activity / Workflow Diagram

Export: 01_ActivityDiagram_LogFeeding.pdf

Capture feeding flow with:

- Required field checks
- Invalid number handling
- Future timestamp preventing
- Save confirmation
- Feeding History list update
- Offline retry path (if included in your scenario)

Goal: Show process thinking before writing stories.

Part 2 — User Stories + Acceptance Criteria (G/W/T)

Export: 02_Stories_AC_LogFeeding.pdf

Write 3–4 user stories:

- Log feeding event ☑
- View list of feedings ☑
- Validate amount ☑
- Handle missing data ✕

(optional: offline save + retry)

Acceptance Criteria must use: GIVEN • WHEN • THEN and Include success + error scenarios based on the workflow.

Part 3 — Low-Fidelity Wireframes (Annotated)

Export: 03_LowFiWireframes_LogFeeding.pdf

Screens required:

1. Add Feeding
2. Feeding History (Empty & Populated)
3. Confirmation state / toast

Annotations must show:

- Required fields + inline errors
- Button disabled state
- Confirmation & navigation behavior

Tools allowed: Figma, Miro, PowerPoint, Canva, or paper → photo

Part 4 — Test Case Matrix

Export: 04_TestCases_LogFeeding.csv

Minimum 8–10 cases, covering:

Category	Example
✓ Success	Valid food entry saved & displayed
✓ Validation	Incorrect units / negative numbers
✗ Time error	Future timestamp blockedd
⟳ Retry	Offline → save + reattempt

UI Verification Button disabled until valid

Columns required:

- ID
- Preconditions
- Steps
- Expected Results

- Status (Pass/Fail once tested)

Part 5 — UAT Walkthrough Screens

Export: 05_UAT_LogFeeding.pdf

Demonstrate:

- ✔ Add new feeding
- ✔ Confirm saved
- ✔ Verify on Feeding History screen
- ✔ Test error message (at least 1)

Use prototype screenshots if real app isn't available.

Audience: stakeholders – use friendly commentary.

Part 6 — Sprint Review Mini-Script

Export: 06_SprintReview_LogFeeding.md

60–90 seconds:

- What we delivered ✅
- How it improves the product 🎯
- Key learning from Sprint 1 applied 🔁
- What's next for Feeding (sorting, editing, nutrition insights)

Tone: confident, concise, user value first.

Part 7 — Portfolio README (Updated)

Export: 07_Portfolio_README_Sprint2.md

Add:

- Before/after from Sprint 1 learnings
- Workflow + wireframe thumbnails

- Test evidence screenshot
- What you contributed

CHAPTER SUMMARY

In **Sprint 2: Feeding Activity**, Team Vertex transformed a fundamental idea of logging pet meals into a functional, polished product increment under real-world pressures. Along the way, they encountered the hidden complexity behind "simple" features: needing to handle input validation, user feedback, and edge cases that only surfaced through diligent testing.

They learned to **inspect and adapt on the fly**, whether it was refining acceptance criteria when Malik exposed a gap, or re-negotiating scope when a stakeholder tried to introduce mid-Sprint change. Despite these challenges, the team kept their eyes on the prize: the Sprint Goal was achieved, and the Feeding Log feature was delivered on time.

Crucially, the Sprint wasn't just about *what* they built (the product), but also about *how* they built it (the process). Team Vertex improved its Agile practice by:

- Strengthening cross-functional collaboration illustrated in Jenna's design handoff meeting and the daily Scrums that swiftly resolved issues.
- Upholding Scrum values in tough moments by saying "no" (or "not now") to a new request to honor their commitment, but doing so transparently and respectfully.
- Embracing a quality-first mindset evident in how Malik's thorough testing and the subsequent fixes were treated as integral to "done," not as afterthoughts.
- Embedding continuous improvement by turning retrospective insights into concrete actions (like the validation checklist and coding standards) that would benefit future Sprints.

By Sprint's end, Team Vertex had not only delivered a valuable new **Feeding Activity** feature (much to stakeholders' delight), but also leveled up as a team. They moved a step further out of early turbulence, finding their groove and establishing norms that would make them more efficient and resilient. Sprint 2 demonstrated that even under pressure, a well-coordinated Agile team can adapt and deliver. The pet owners using PetTracker now had a convenient way to track feedings, and Team Vertex had a stronger foundation in both product and process for the challenges ahead.

WHAT'S NEXT?

With pets now munching under watchful logging, Team Vertex sets its sights on **Sprint 3**, where a new star feature awaits: *Walk Activity Management* and the *Pet Profile Sharing*. The feature, eagerly requested by Marketing, promises to let users share their pet's profile with friends, an exciting leap into the app's social potential. The team will carry forward Sprint 2's lessons as they tackle this larger, more interactive story. Will the groundwork laid (clear definitions, checklists, reusable patterns) help them accelerate development, or will new complexities emerge as they venture into sharing and social functionality? As new expectations and perhaps new team dynamics come into play, one thing is certain: Team Vertex's journey of learning is far from over. **In Sprint 3, how will they balance creative innovation with disciplined execution?**

You never know, the next Sprint might bring the team into uncharted territory, with all-new challenges to conquer and lessons to learn.

Feeding Log Feature Breakdown

User Story	Description	Primary Acceptance Criteria
US-01: Basic Entry	As a pet parent, I want to enter the food name and portion size so I can track what my pet eats.	✅ Food name and portion size fields are required ✅ Portion must be a positive number
US-02: Error Handling	As a user, I want to be alerted when I leave a field blank or enter invalid data.	✅ Blank inputs trigger inline validation ✅ Invalid entries (e.g., emoji, text in portion field) are rejected
US-03: Confirmation Feedback	As a user, I want to see a confirmation when my food log is saved so I feel confident it's recorded.	✅ Success message appears after valid submission ✅ Form clears automatically
US-04: Presets (Future Sprint)	As a user, I want to select from a list of common food presets to save time.	⏳ To be refined in future release (non-MVP)

SPRINT 3

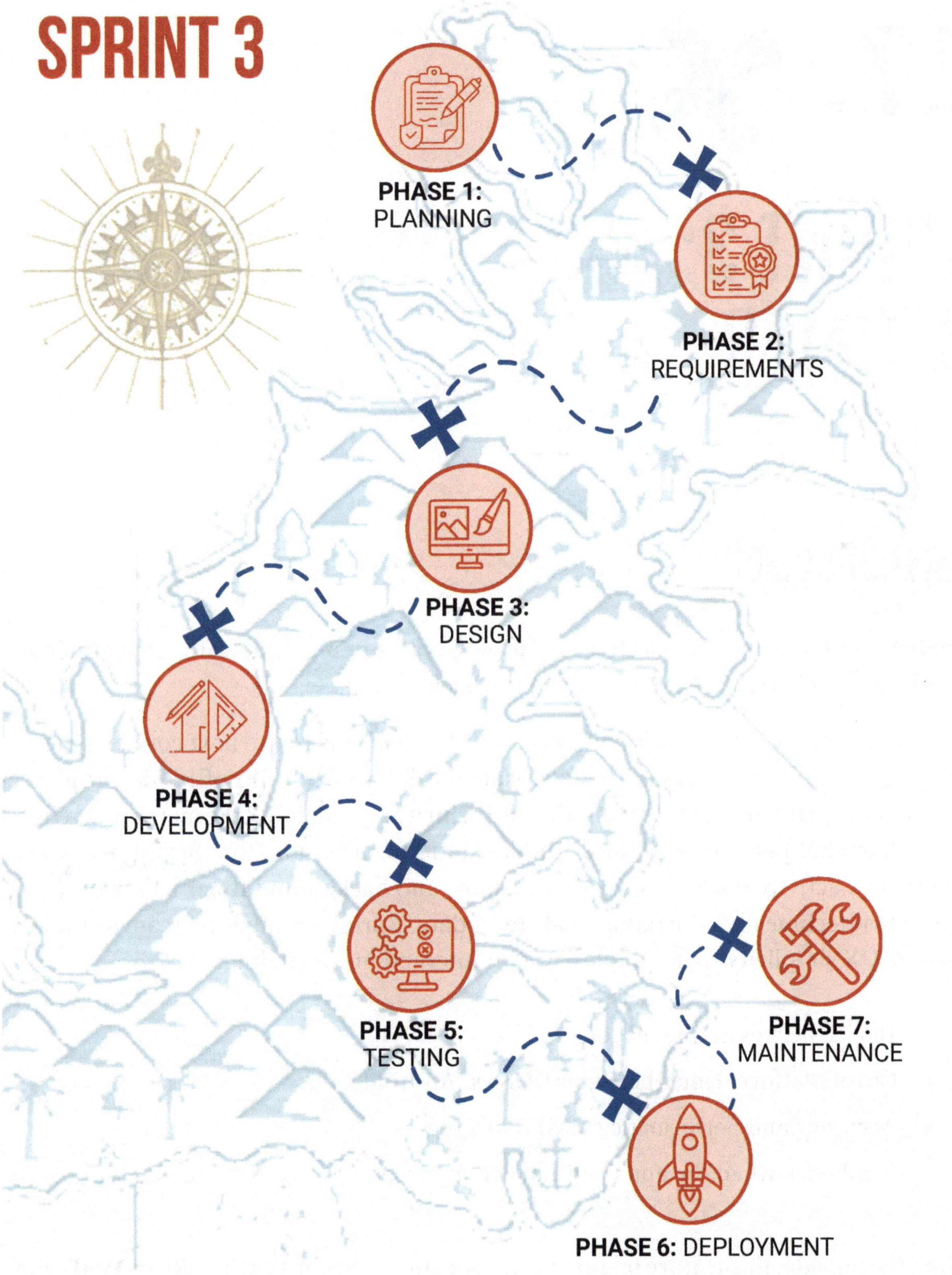
PHASE 1:
PLANNING
PHASE 2:
REQUIREMENTS
PHASE 3:
DESIGN
PHASE 4:
DEVELOPMENT
PHASE 5:
TESTING
PHASE 6: DEPLOYMENT
PHASE 7:
MAINTENANCE

17. SPRINT 3: WALK LOG WITH GPS & TIMER

INTRODUCTION

Sprint Goal: Enable users to track their pet's walks in real time with a live timer and GPS logging, even when the app runs in the background.

With food logging complete, Team Vertex shifted its focus to the next core experience: tracking pet walks in real time. They also slate the deferred **"Pet Profile Sharing"** user story (from Sprint 2) for implementation in Sprint 3, eager to fulfill the marketing request for a shareable pet profile feature. Unlike static form inputs, this Sprint introduced continuous activity tracking, GPS permissions, and background timer behavior. It was their most technically demanding feature to date. Sprint 3 tested the team's ability to balance UX simplicity with complex system interactions, including:

- Device permission management
- Cross-platform timer behavior (iOS vs. Android)
- Accuracy and performance of GPS logging
- Resilience when the app is minimized or closed

From the outside, this feature might appear as a simple **Start Walk → Stop Walk** flow, but for Team Vertex, it revealed deep differences in platform behavior, required proactive debugging, and demanded real-time collaboration across design, development, and QA.

This Sprint not only delivered an exciting new user-facing feature, but it also matured the team's engineering discipline and strengthened their coordination under pressure.

OBJECTIVE

Sprint 3 marks a leap in complexity that required tight coordination and deep architectural alignment by the team. The goal was to deliver a real-time walk tracking feature (with live timer and GPS logging) while maintaining Agile discipline under pressure. In doing so, Team Vertex proved they could build complex, real-time functionality while staying adaptable and aligned as a team.

FEATURE IN FOCUS: WALK ACTIVITY MANAGEMENT

Walk Log with GPS & Timer: This Sprint's feature enables pet owners to log their dog walks with a live timer and automatic GPS tracking. The app records the duration and route of each walk in real time, even if the user switches apps or the phone is locked. This capability transforms PetTracker beyond static data entry, letting users effectively "schedule" a walk and have the app track it for them in the background. A gentle notification reminder even prompts the user to resume tracking if movement stops unexpectedly during a walk. It's a core quality-of-life feature that keeps users engaged and makes tracking pet wellness effortless.

TEAM VERTEX IN ACTION

Just before Sprint 3 kicked off, Maria (Product Owner) led with a focused and technically ambitious goal:

"Let's give users a real-time walk tracker that keeps logging even when the app's in the background."

This introduced a leap in complexity for Team Vertex: timers, GPS integration, platform-level permission handling, and session continuity. The solution would require tight coordination and deep architectural alignment.

To get ahead of the unknowns, the team held a structured (JAD) session before planning. Maria facilitated the discussion, which included:

Workflow Mapping: Jenna (UX Designer) and Malik (QA) diagrammed the entire Walk Log flow, from Start to Save, across both iOS and Android.

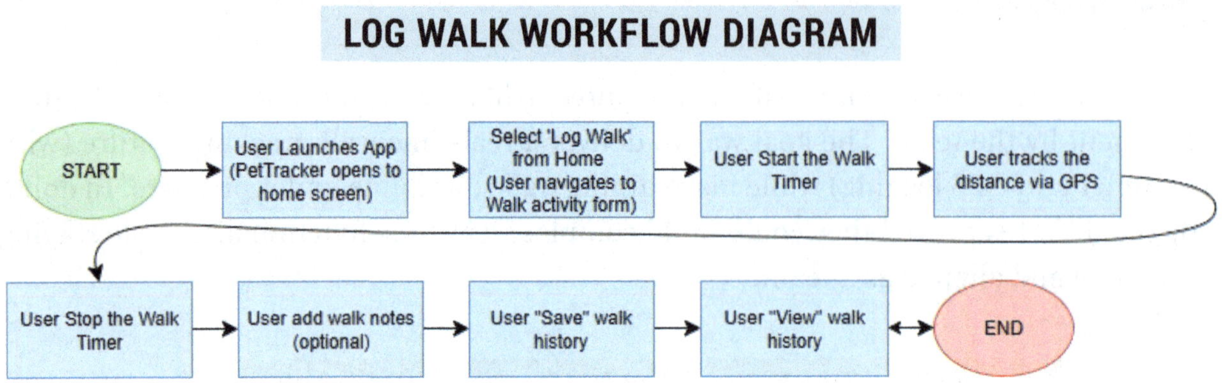

Role Clarification:

- Maria: Schema design and story grooming
- Jenna: Timer screen UI with states
- Alex: Timer logic and GPS integration
- Malik: Edge case scenarios and test design

Key Questions Answered:

- How does the timer persist during backgrounding?
- What happens if GPS is denied mid-walk?
- When does data get saved: at Start, Stop, or Save?

(The team even converted a whiteboard sketch into a workflow diagram attached to the Jira Epic. This visual became a reference point throughout Sprint 3.)

REQUIREMENTS: BREAKING DOWN THE FEATURE

Turning Features into User Stories: In Sprint 3, Team Vertex moved from concept to clarity by breaking the Walk Log feature into three focused user stories:

- **Start Walk:** As a user, I want to tap "Start Walk" so that the app begins tracking time and GPS location.
- **Stop Walk:** As a user, I want to tap "Stop Walk" to end the session and save the walk data.
- **Save Walk:** As a user, I want to see a success message and have my walk saved for review.

These stories helped the team isolate functionality and plan development tasks in manageable, testable units.

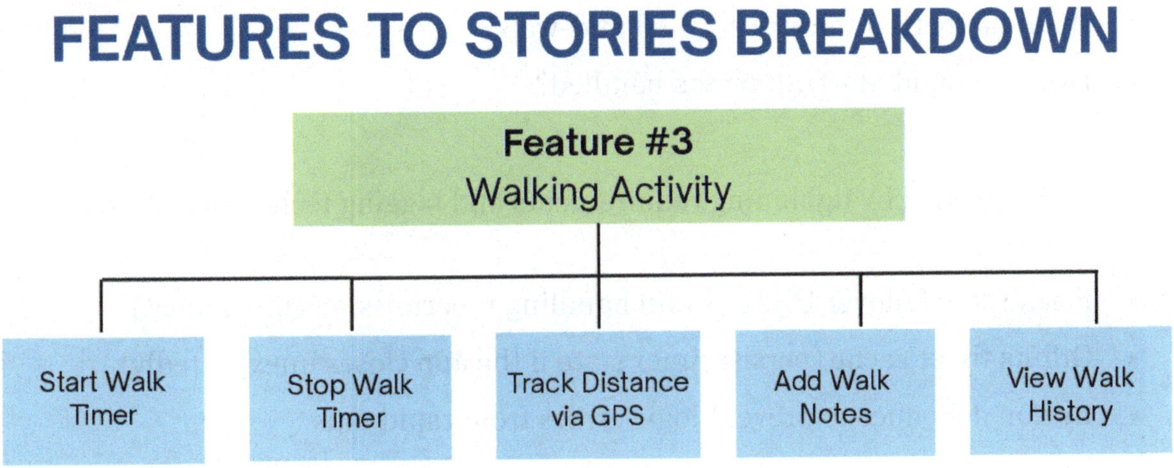

Defining Acceptance Criteria: To ensure alignment and build confidence in the Walk Log behavior, Maria drafted detailed acceptance criteria for each story. The team reviewed and refined the following:

Start Walk:

- App must request GPS permissions on first use
- Timer starts at 00:00 and increments every second
- Location logs begin in a background thread

Stop Walk:

- Timer stops and freezes on the final time
- Location logging ends
- Walk summary screen displays total duration

Save Walk:

- Walk is stored locally and synced to the server
- A confirmation message is displayed
- Data includes timestamp, duration, and basic route (start and end points)

During refinement, Malik raised critical concerns about **Edge Cases & Constraints:**

- What if the user denies GPS access?
- What happens if the app is killed mid-walk?
- How are rapid Start/Stop taps handled?

The team responded by tightening requirements and tagging technical tasks for:

- Permission failover UX (graceful handling if permissions are denied)
- Offline timer cache (persist timer state if the app closes unexpectedly)
- Button debounce to prevent logic errors from rapid taps

LOG WALK ACTIVITY DIAGRAM

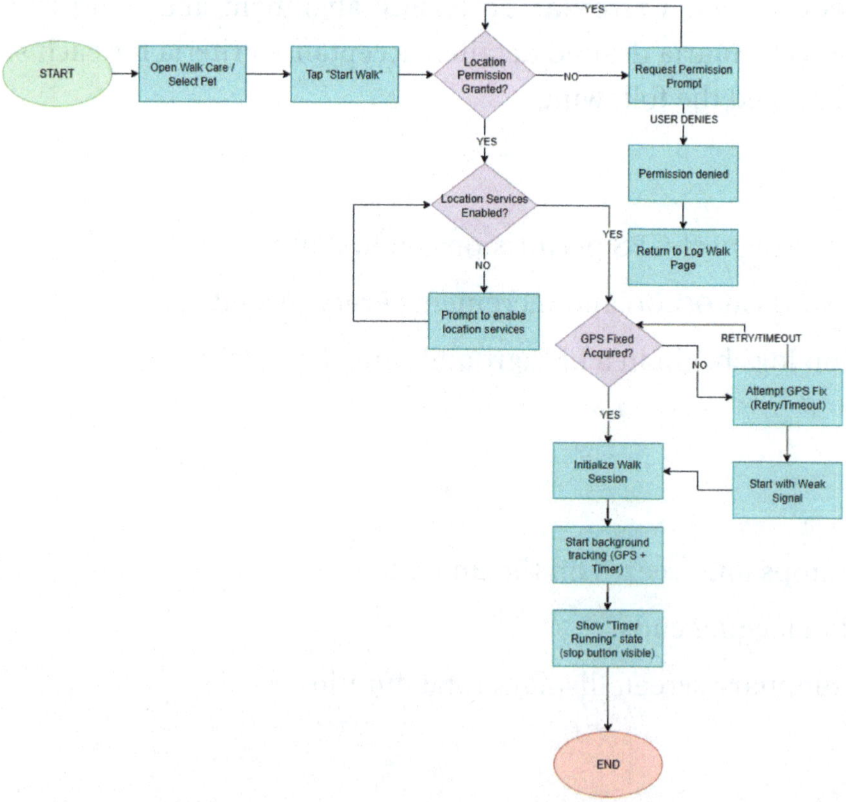

Here's a concise walkthrough of the **Start Walk Timer – UML Activity Diagram**.

Purpose

Shows the control flow to start a walk session safely with GPS, from user tap to a running timer.

Preconditions

- App is open on Walk Care; a pet is selected.

Main Flow

1. **Initial node → Open Walk Care / Select Pet → Tap "Start Walk."**
2. **Check Location Permission.**
   - If granted → proceed.
   - If not granted → show **Request Permission Prompt**; return to the check after user responds.
   - If explicitly **Denied** → end this attempt with guidance to retry later.
3. **Check Location Services (OS setting).**
   - If enabled → proceed.
   - If disabled → **Prompt to Enable Location Services**. If user declines, flow ends (User Cancels/Returns).
4. **Acquire GPS Fix.**
   - If fix obtained → **Initialize Walk Session** (timer=0, distance=0, route=[]).
   - If not, **Attempt GPS Fix (Retry/Timeout)** loop. On timeout, offer **Start with Weak Signal (Optional)**. Choosing it continues to initialization.
5. **Start Background Tracking** (GPS + timer).
   - Show **"Timer Running" UI** with Stop control → **Final node** for this start procedure.

Alternate/Exception Paths

- **Permission denied**: Explain why needed; user must change app permission to proceed.

- **Location services off**: User must enable OS-level location; otherwise exit.
- **No GPS fix**: Retry loop; optional weak-signal start to avoid blocking.

Postconditions

- A walk session is active; timer running, GPS tracking started, UI updated to "running."

They also captured **Documentation & Artifacts:** all user stories were linked to corresponding Figma mockups; edge case handling was documented in Confluence; and Jira tickets included test notes, a device testing matrix, and clear component ownership. This requirements phase ensured Sprint 3's technical complexity didn't become functional confusion. By writing sharp stories and defining solid guardrails, Team Vertex stayed aligned *and* ready to build.

Visualizing Movement: Designing the Walk Log UI. Designing the Walk Log screen presented unique challenges. Unlike previous screens, the Walk Log wasn't a simple form: it required continuous interaction, real-time feedback, and accessible controls on the go. Jenna approached the design with clarity and motion in mind, after the team had aligned during Sprint Planning on what the feature needed to deliver.

Sprint Goal Breakdown:
- **Primary Objective:** Build a Start/Stop walk experience with a real-time timer and basic GPS tracking.
- **Success Criteria:**

☑ Timer persists in background
☑ GPS records location at intervals
☑ Walk entry is stored with timestamp, duration, and notes (optional)

Spike Review:
Two key Sprint 0 explorations were revisited:
- *Timer Behavior:* variability in timer thread handling across platforms
- *GPS Feasibility:* Expo plugin vs. native SDK trade-offs, and battery impact

- *Team consensus:* Use Expo's Background Timer, with a fallback caching system for any unsupported behavior.

Sprint Planning Highlights:

Stories Committed:

- Timer with start/stop logic: *8 points*
- Background tracking with permission flow: *13 points*

Execution Plan:

- UI design due by Day 2 (Jenna)
- Paired development on timer/GPS logic (Alex & Malik)
- Schema design & validation (Maria)
- QA scenarios for force-close, denied permissions, GPS pause (Malik)

Risks Identified:

- Android location permission quirks
- Need for GPS simulation tools
- Cross-platform differences in background services

Readiness Checklist:

- Walk Log workflow diagram created and uploaded
- Figma mockups linked to user stories
- Definition of Ready confirmed for all stories
- Test plan and edge case coverage identified
- Story points reviewed against team capacity (they ensure the commitment is realistic for the 2-week Sprint, considering one big feature plus possibly Pet Profile Sharing from backlog if time permits).

With assumptions challenged and workflows mapped visually, Team Vertex began Sprint 3 with confidence, alignment, and an elevated understanding of the system they were about to build both technically and collaboratively.

Design Planning Highlights

In the Sprint Planning session, Maya (Scrum Master) underscored the feature's complexity and guided the team through story sizing and task allocation. Key stories discussed from a design perspective included:

- Start/Stop Timer UI with real-time tracking
- Background behavior and GPS permission handling
- Notes field for post-walk input
- Save confirmation and data submission flow

Jenna committed to finalizing the high-fidelity mockups by Day 2 to unblock development. Maria ensured all acceptance criteria were reflected in Jira, and Malik flagged the need for clearly defined UI states to support thorough QA testing.

Key Mockups: Jenna's Figma designs for Sprint 3 included two main screens:

- **Walk Timer Screen:**
  - Large digital timer (MM:SS) centered on screen
  - Prominent **Start**, **Stop**, and **Save** buttons
  - Subtle walking-icon animations to indicate tracking status
  - Color-coded states (green = active, red = stopped, etc.) to show timer status
- **Post-Walk Summary:**
  - Optional **Notes** field for context (e.g., "Morning walk before breakfast")
  - Total walk duration displayed
  - **Save Walk** confirmation button
  - Success checkmark modal after saving a walk

LOG WALK TIMER MOCKUP

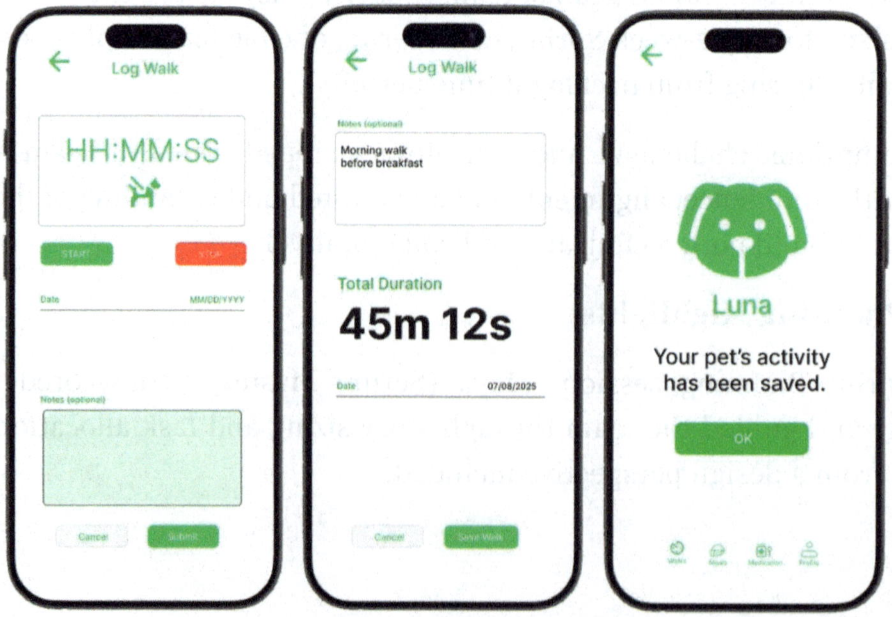

Accessibility & Usability Enhancements: The design accounted for real-world usage. Jenna ensured tap targets exceeded the minimum size for easy use while walking outdoors. Color contrast was checked for readability in sunlight. VoiceOver/TalkBack screen reader support was mapped for the timer, buttons, and notes field. Visual state indicators (icons + text) confirm when tracking is active or paused, so users always know if the app is logging their walk.

Design Iterations: After presenting the initial mockups during a mid-Sprint design review, the team made a few quick tweaks based on feedback:

- Maria requested that the Notes field be optional and collapsed by default (to reduce screen clutter).
- Alex proposed using floating action buttons for Start/Stop to prevent thumb-reach issues on larger phones.
- Malik flagged potential confusion between the **Stop** and **Save** buttons, prompting Jenna to refine the icons and add spacing to differentiate them.
- Jenna rapidly applied these updates and marked the designs with a "Design Complete" tag in Jira.

Outcome: By the end of the Design phase, all mockups were finalized and approved in Figma. Design tickets were updated and linked to the relevant user stories. Accessibility considerations were fully integrated. Most importantly, the team's alignment from planning carried through to the designs, with the visuals faithfully reflected the agreed requirements. This strong design foundation paved the way for seamless development and testing execution.

Building Time into Motion: Development

Sprint 3 marked a major step up in technical complexity. Static screens gave way to live timers, background threads, and device-level permissions. Alex (Developer) led the implementation effort with a focus on accuracy, responsiveness, and consistency across platforms.

Each morning during Daily Scrum, blockers were surfaced early. For example, Alex flagged discrepancies in GPS permission behavior between iOS and Android, while Malik raised concerns about slight timer drift on some Android devices. These daily syncs helped the team course-correct quickly without derailing momentum.

Timer Implementation: Alex built the core walk timer with two key requirements in mind:

- **Real-Time Increments:** The on-screen timer should display time in **MM:SS** format, updating every second.
- **Background Continuity:** The timer must not reset or pause when the app is minimized. Alex leveraged native device APIs and background services to keep the timer running even if the app was in the background. He also implemented a fallback caching system to snapshot the time in case the app was closed or the OS reclaimed resources, so the timer could recover its state on return.

GPS Integration: Next, Alex integrated GPS location tracking using platform-specific APIs:

- **First-Time Use Prompt:** The app triggers the native GPS permission modal on first use, handling both "Allow While Using App" and "Deny" flows gracefully.
- **Active Walk Logging:** Once running, the app captures a geolocation "ping" every 10 seconds and stores the coordinates (focusing on start and end points of the walk for now).
- **Battery-Aware Design:** The tracking auto-pauses if no movement is detected for 60 seconds to conserve battery. It even sends a gentle reminder notification to the user to resume tracking when activity picks up again.

LOG WALK GPS INTEGRATION

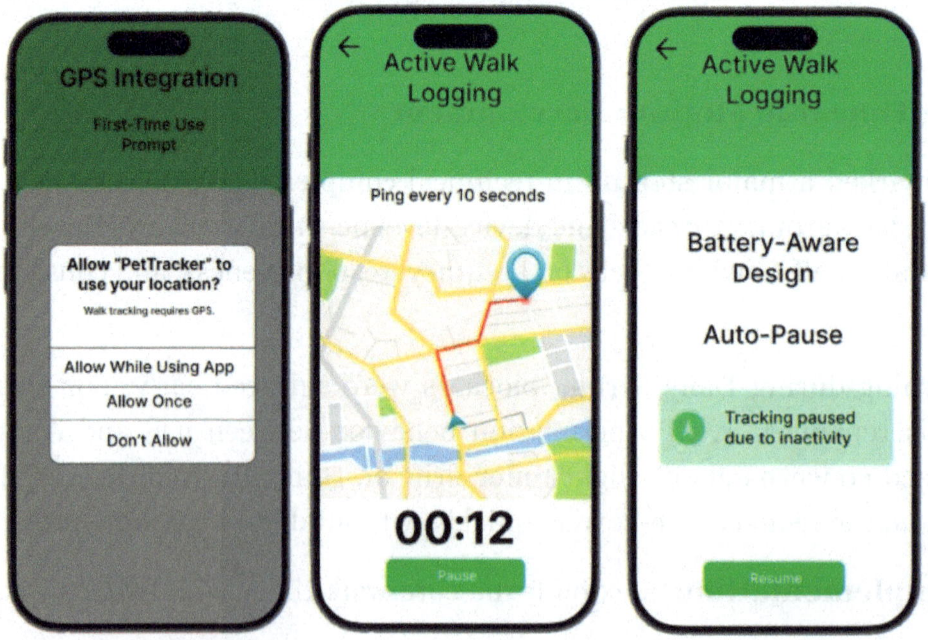

Permission Handling: Malik (QA) and Maria teamed up with Alex to map out conditional behaviors for various permission scenarios:

- **Denied permissions** → fallback to timer-only tracking (if location permission is never granted, the app still logs time)
- **Revoked mid-walk** → user is alerted with guidance to re-enable permissions in app settings
- **Incomplete permission state** → soft-block the walk start with an option to retry, plus help text explaining why GPS is needed

These logic branches were reviewed in real time during Scrum to ensure no edge case was overlooked.

XP PRACTICE #3 – REFACTORING FOR RESILIENCE

After implementing the timer, background recovery, and GPS flows, Alex reviewed the code with a critical eye. The logic worked — but it was growing dense, hard to follow, and error-prone. Rather than defer cleanup, the team embraced their third XP discipline:

Refactoring: clean code, even under pressure.

- Alex split large blocks into modular helper functions
- Permission handling logic was extracted into its own reusable utility
- The timer's caching and recovery routines were simplified into a single interface
- Unit tests were added after refactoring to validate behavior hadn't changed

The team treated this as a non-negotiable part of the development rhythm, not a luxury. Why? Because velocity without clarity is a trap. And unreadable code today becomes broken code tomorrow.

🔁 **Maya reinforced that refactoring is:**

- Part of the Sprint, not outside of it
- Included in story estimation
- Enforced via the Definition of Done

The outcome? Fewer bugs. Faster debugging. Happier developers.

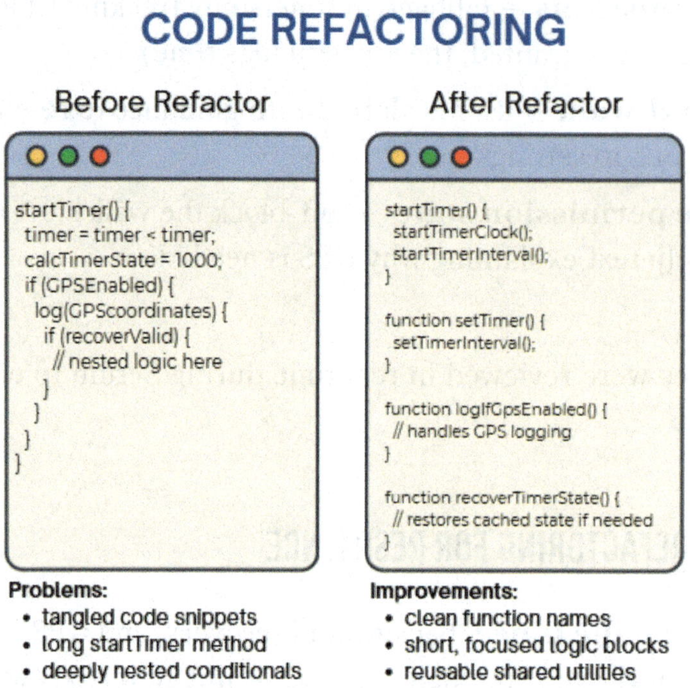

Pairing & Review: To strengthen delivery quality across both platforms, the team embraced pair programming and code review:

- Alex pair-programmed the Android portion with Malik, aligning development with test cases from the start.
- Jenna validated the iOS UI on multiple screen sizes and with accessibility settings enabled, to catch any layout issues early.
- Maria reviewed the backend data payload structure to ensure the stored walk records would be complete and compatible with future reporting features.

Team Vertex used every feedback loop available, from code reviews to daily stand-ups, to manage risk and ensure clarity. This development phase proved they could build complex, real-time features while staying Agile under pressure.

Testing Time in Motion

Sprint 3's **Testing** demanded more than typical UI validation; it tested Team Vertex's ability to simulate dynamic, real-world user behavior. Malik took the lead on QA, focusing on timing precision, permission handling, and background stability.

One major issue emerged during exploratory testing: **The Timer Reset Bug.** Malik identified a critical flaw:

- **Bug:** If the app was force-closed mid-walk, the timer reset to 00:00 on reopen, causing data loss and a confusing user experience (it was as if the walk had just started).
- **Root cause:** The timer relied solely on in-memory state. No walk data persisted if the app was killed, so the session information vanished on restart.

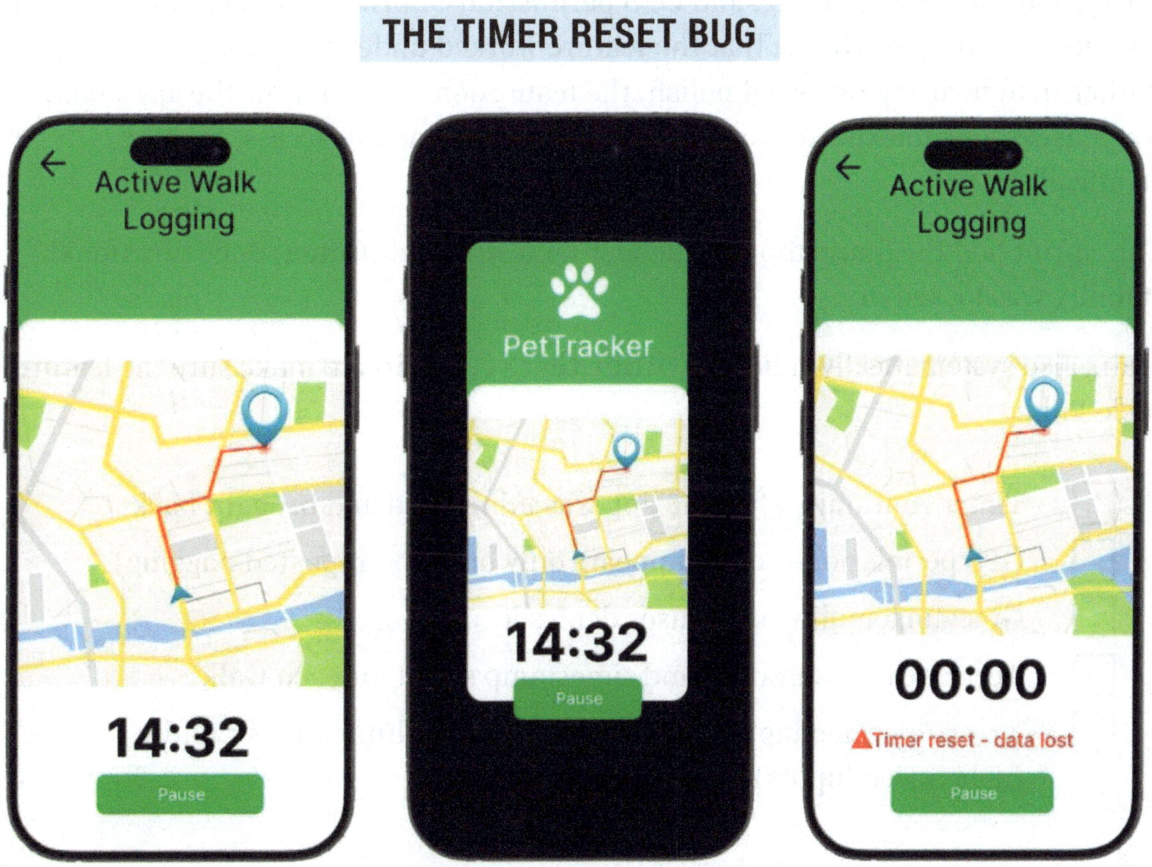

Resolution: Alex tackled this immediately by implementing a more robust state persistence:

- Storing the walk's start timestamp in persistent local storage as soon as a walk begins.
- On app reopen, checking if any active walk session exists (using that stored

timestamp) and recalculating the elapsed time so the timer resumes where it left off.

- Blocking any duplicate "Start" actions if a session is already in progress (to prevent multiple overlapping timers).

This solution ensured users could trust the app to remember an ongoing walk, even after an accidental app closure.

Layering Quality: Why Sprint 3 Focused on Functional Testing

With the Walk Log feature introducing complex behaviors like real-time timers, background session handling, and GPS permissions, Sprint 3's QA efforts prioritized functional testing—verifying that the feature worked under real-world conditions. Rather than focusing on visual polish, the team zoomed in on how the app behaved when users took unexpected actions (force closing, denying access, tapping erratically).

This sprint laid the foundation for deeper UI testing in Sprint 4, once functional stability was locked in.

Malik also systematically validated **other test scenarios** to make sure the feature was rock-solid:

- [ ] Timer continuity after the app was minimized and brought back
- [ ] GPS permission prompt appears only once (no repeated nagging)
- [ ] Graceful handling if the user denies location access
- [ ] Accurate total duration and timestamp saved for each walk
- [ ] No errors when tapping Start/Stop/Save multiple times rapidly (debounced inputs)

To support ongoing quality, Malik created lasting **QA artifacts** for the project:

- He added specific walk timer and GPS-retry test cases to the team's reusable QA suite in Confluence (so future Sprints could reuse these tests).
- Created a device matrix of test cases to cover different phones (iOS and Android variants, including low-end devices to catch issues like the timer drift).

- Documented the permission handling edge cases and the user alert messages for each scenario.

By the end of testing, the team had confidence that the Walk Log feature met all criteria and handled adversity (like lost GPS or app closures) gracefully.

UAT DEPLOYMENT: WALK LOG IN ACTION

Scene: Day 8 of Sprint 3. Development and QA are complete, and Team Vertex prepares for **Deployment** to the UAT (User Acceptance Testing) environment:

Maya (Scrum Master): "Alright, team, we've got QA sign-off. Alex, are we good to kick off the UAT deployment for the Walk Log feature?"

Alex (Developer): "Yep, all merge checks are green. I just pushed the final branch into develop, and the CI pipeline has started running the UAT build job."

Malik (QA): "I'll prep a smoke-test checklist. We need to confirm the GPS lock, timer accuracy in the background, and data saving on both platforms."

Maria (Product Owner): "Once it's deployed, I'll notify stakeholders that the UAT app link is live and ready for testing. Please confirm once it's up."

(A few minutes pass as the CI/CD pipeline deploys the new build.)

Alex: "Deployment to UAT is complete. The build passed all CI checks and was published to our UAT environment. Here's the link: ***UAT Walk Log Build***."

Malik: "Opening it now... okay, on iOS: tapping **Start Walk** triggers the timer... backgrounding the app... and back. Nice, it's still counting! I'll run the same flow on Android next."

Jenna (UX Designer): "Keep an eye on the GPS icon placement on Android. The prototype had a persistent indicator in the status bar. I want to confirm it's showing up correctly under actual device conditions."

Malik: "Confirmed on both platforms. Timer persists, GPS is active, and the Save flow returns a success message. Now I'll test some edge cases: dropping GPS signal, switching apps mid-walk, low battery mode, etc."

Maria: "Great. I'm messaging stakeholders now and tagging our demo steps in Confluence. Let's be prepared to walk them through this flow in tomorrow's review meeting."

Maya: "Great job, team. I'll update the Jira board to mark **UAT Deployment: Complete**. Be sure to note any issues during the stakeholder walkthrough so we can address them either as hotfixes or backlog items."

Shipping the Experience: With development complete and QA sign-off secured, the Walk Log feature was deployed to UAT. This was Team Vertex's first deployment involving live timers, background behavior, and GPS integration, all requiring coordinated validation across iOS and Android. Alex led the deployment using the CI/CD pipeline set up in Sprint 0. After a successful build and smoke test, the feature was accessible via the shared UAT app link for internal demo and stakeholder review.

During the **Sprint Review** on Day 9, the team demonstrated the new feature live. Alex launched the UAT app and initiated a walk: he tapped **Start Walk**, then immediately minimized the app. After a few seconds, reopening the app revealed the timer had **continued counting in the background**. He then tapped **Stop Walk**, entered a quick note, and hit **Save Walk**. A confirmation message appeared, and the walk record was successfully stored. Stakeholders watched this demo in real time on both an iPhone and an Android device, and they were impressed by the smooth experience and consistent data.

Stakeholder Feedback: The review yielded positive feedback and a couple of great ideas. One stakeholder suggested adding a more visible UI cue to indicate GPS tracking is active while the app is minimized (so users are reassured it's working). Another proposed a future enhancement: a **Walk History** screen so pet owners can track their weekly exercise trends. Maria captured these suggestions and logged them in Jira, tagging them for consideration in the product backlog (they might even become Sprint 4's goal!).

Importantly, **Team Vertex also delivers the Pet Profile Sharing feature during Sprint 3**, fulfilling the promise made in the last Sprint. They demo how a user can tap a "Share Profile" button on the pet's profile screen to generate a cute card with the pet's photo and details, which can be shared via social media or messaging. This smaller story was completed alongside the Walk Log work. The stakeholders from Marketing smile at this: they got their wish, and it's ready in time for the upcoming trade show. The team explains that by deferring it to Sprint 3, they were able to give it proper attention without compromising the Feeding Log earlier.

By deploying the Walk Log feature to UAT and walking through a real-world demo, Team Vertex validated that their solution not only worked but worked reliably **in motion**. The successful Sprint reaffirmed the team's ability to handle time-sensitive, location-aware features with confidence.

AGILE IN MOTION

Sprint 3 showcased several Agile and SDLC best practices in action:

- **Upfront Design Collaboration:** Before development, the team conducted a JAD session to collectively sketch out the Walk Log workflow. This early collaboration (with a shared diagram artifact) gave everyone a clear vision and aligned expectations, embodying *"planning is everything"* in Agile.
- **User Story Refinement:** The feature was broken into well-scoped user stories with clear **Definition of Ready** (DoR). Maria ensured each story had detailed **acceptance criteria** and that UX mockups were attached in Jira, so nothing was ambiguous when development began.
- **Continuous Integration & Small Increments:** Rather than a big bang approach, Alex and Malik worked in small increments (e.g., first get the timer working, then integrate GPS, then add persistence) and integrated their code frequently. Every merge underwent automated CI checks, reflecting an Agile approach to constant integration and feedback.
- **Daily Stand-ups & Rapid Iteration:** Daily Scrums were used to surface blockers (like the cross-platform discrepancies) early, allowing the team to fix issues or adjust the approach the very same day. This iterative problem-solving

kept the Sprint on track despite its technical complexity.

- **XP (Extreme Programming) Practices:** The team practiced pair programming and code review (developer + QA pairing on implementation, designer verifying UI in dev builds, PO reviewing data schemas). This reduced handoff gaps and ensured quality was built in. It also reflects the *"whole team approach"* in Agile, where quality and outcomes are everyone's responsibility.
- **Risk Management:** Potential risks (permission quirks, battery impact, etc.) were identified during planning and tracked. The team implemented fallbacks (like offline caching for the timer, and a debounce to prevent double-taps) preemptively. This proactive risk mitigation is part of Agile's adaptive planning, addressing "known unknowns" early.
- **Definition of Done (DoD):** Team Vertex's DoD was effectively expanded this Sprint. They didn't consider a story done until it met all AC, passed regression tests (Malik's extensive scenario testing), and had any new test cases documented in Confluence. The inclusion of tasks like updating the Mobile QA checklist and Confluence docs as part of the Sprint work shows their growing emphasis on *sustainable development* (maintaining documentation and test suites for the future).
- **Retrospective & Adaptation:** Finally, the Sprint ended with a thorough retrospective where the team celebrated successes and identified improvements (like better handling of iOS permission oddities). They immediately turned these into concrete action items for the next Sprint. This continuous improvement loop is a hallmark of Agile teams moving toward high performance.

Throughout Sprint 3, Team Vertex demonstrated agility not just in name but in practice, refining requirements collaboratively, delivering in small slices, and continually reflecting and improving.

TEAM REFLECTION

Lessons That Stick: After deploying their most technical feature to UAT, Team Vertex took time to reflect in the Sprint 3 Retrospective. They celebrated the team's success across both platforms and discussed how to handle mobile inconsistencies more proactively going forward. Jenna noted that iOS handles permission prompts differently than Android (for example, if a user previously denied location access, iOS might suppress the prompt entirely on subsequent tries). Malik observed that background timers drifted slightly on lower-end Android devices. These nuances, while resolved during development, highlighted the need for better knowledge sharing and platform-specific testing strategies in the future.

To support future Sprints and new team members, they decided to enhance their project documentation in Confluence. Malik documented the known behavioral differences between Android and iOS regarding background tracking, timer persistence, and permission handling. Maria created a **"Permission-Heavy Stories" checklist** to ensure that any future user story involving system permissions will include things, like: prompt logic, denial fallback messaging, retry flows, and a state-persistence strategy. These additions became part of the team's growing **Mobile QA Toolkit** in Confluence, serving as reusable assets for upcoming features (such as walk history, notification reminders, or location-based tagging).

Scene: Sprint 3 Retrospective Meeting (Day 10, late afternoon). The team gathers in their hybrid format: some in the office, others on video. Maya has set up a virtual whiteboard with three columns: *What Worked*, *What Didn't*, and *What We'll Try*. The retrospective conversation begins:

Maya (Scrum Master): "Alright, Sprint 3 was a big one. Let's start with what worked well. Who wants to kick us off?"

Jenna (UX Designer): "I'll start. Honestly, seeing the timer persist across background states on both iOS and Android was a huge win. It reflected the design exactly as intended, and I was relieved it translated so smoothly into the real app."

Malik (QA): "Agreed. Our QA coverage held strong despite all the edge cases we tackled. Though I did notice a small drift in the timer on some low-end Android phones: a few seconds off after resuming from background. We patched it in Sprint, but it caught me off guard initially."

Alex (Developer): "That drift was due to how those devices aggressively throttle background processes. I think we handled it well once we spotted it, but I probably could've flagged that risk earlier during planning."

Maya: "Great insight. Now, what didn't go as smoothly?"

Jenna: "I ran into inconsistent permission behavior, especially on iOS. If a user had denied location access earlier, the app just... didn't prompt again at all. We hadn't designed a fallback screen for that scenario, so it was a gap."

Malik: "That's something I'll add to our Mobile QA Toolkit documentation: how to handle system-level permission flows and differences. I've actually started a Confluence page outlining the platform differences so we don't forget."

Maria (PO): "I also created a checklist template for any permission-heavy user stories going forward. It covers things like providing a fallback message if the app can't prompt, a retry option, and persisting the user's intent. I'll attach it to our user story template so we remember to consider these."

They briefly touch on the Pet Profile Sharing feature: how deferring it was wise. Because when they did implement it in Sprint 3, they found a couple of UX things to tweak (which they managed within the Sprint). Had they forced it into Sprint 2, it might have been a rush job. The lesson: trust your capacity and don't be afraid to say "next Sprint" to even cool ideas, to do them well. Marketing is happy now because the feature is polished.

After discussing what worked and what didn't, the team agrees on concrete improvements. Maya steers them to wrap up with **action items**:

Maya: "Sounds like we're already addressing some of the gaps. Let's summarize the action items we'll take away from this Sprint."

- Hold platform-specific refinement walkthroughs for features involving GPS, sensors, or any native device APIs (to catch platform differences early).
- Maintain a "permission state" validation matrix for testing scenarios when permissions are granted, denied, or changed, so QA covers all cases.
- Review the mobile behavior checklist (our new checklist for sensor/permission stories) during Sprint Planning as part of our Definition of Ready, so those considerations are never missed.

Maya: "Documenting these learning loops is what makes us stronger. Great work, everyone. Let's carry this momentum into Sprint 4!"

Sprint Retrospective

STOP	START	CONTINUE
Inconsistent permission handling (no fallback screen when denied)	Platform-specific refinement walkthroughs	Cross-functional documentation updates
Reactive risk assessment (flagging background throttle risk earlier)	Review mobile behavior checklist in Sprint Planning	Thorough QA coverage

(From their reflection, Team Vertex committed to all the above improvements, ensuring the hard-won lessons of Sprint 3 wouldn't be forgotten. Sprint 3's Maintenance phase ensured that the knowledge gained from building and deploying the Walk Log feature wouldn't live in siloed minds, but would be captured in a scalable way for the team's future growth.)

CODE REVIEW

1. Push Code & Open PR

A developer pushes code to a shared branch and opens a Pull Request (PR) on a platform such as GitHub.

2. Review Code

Teammates review the code, leave inline comments, suggest improvements, or ask clarifying questions.

3. Approve & Merge

Once feedback is addressed, the PR is approved and merged into the main codebase.

What Reviewers Look for:

Functionality Does the code do what the story requires?	**Readability** Can another dev understand it quickly?	**Tests** Are unit tests included and meaningful?
Reusability Are there duplicate codes? Can it be refactored?	**Error Handling** Are edge cases covered? Are failures handled	**Security** Are inputs validated? Any sensitive data exposed?

Exercises & Activities

Quick Check: Test your understanding of Sprint 3 with a few questions:

1. Why did Team Vertex hold a JAD session before starting Sprint 3, and what did it achieve?
2. What technical challenges made the Walk Log feature more complex than previous features? Name at least two.
3. How did the team ensure the walk timer would continue running if the app was minimized or closed?
4. What was the root cause of the "timer reset" bug, and how was it fixed?
5. In the Sprint 3 retrospective, what was one improvement the team decided to implement for future Sprints?

Apply This: Use Sprint 3's content to practice creating a user flow. Below is an exercise scenario drawn from the team's work:

SPRINT 3 — EXERCISE PACK

Feature: Log Walk Activity

Focus: GPS tracking · Timestamp handling · Data accuracy

Sprint Theme: Movement + metrics — the pet's health picture comes alive

PART 1 — ACTIVITY / WORKFLOW DIAGRAM

Export: 01_ActivityDiagram_LogWalk.pdf

Include:

- Start Walk → Tracking timer begins
- GPS permission decision:

  YES → track route + duration

  NO → fallback manual entry OR deny tracking with explanation
- Pause/Resume option
- End Walk → Save walk entry
- View Walk History (empty + populated states)

Error paths to include:

- GPS denied → alternative input
- Invalid data → validation message

PART 2 — USER STORIES + G/W/T AC

Export: 02_Stories_AC_LogWalk.pdf

Minimum 3 stories:

1. Log Walk with GPS
2. Manual Walk Entry fallback
3. View Walk History

Acceptance Criteria Examples:

- Future timestamps block ✅
- Duration auto-calculated ✅
- Handling denied permissions ✅

PART 3 — LOW-FI WIREFRAMES (ANNOTATED)

Export: 03_LowFiWireframes_LogWalk.pdf

Screens required:

1. Start Walk
2. Tracking in progress (map OR placeholder)
3. Save/End Walk
4. Walk History

Annotations:

- Permission request screens
- Validation messages
- Tracking state indicator (timer, steps, icon)
- Navigation behavior after save

PART 4 — TEST CASE MATRIX

Export: 04_TestCases_LogWalk.csv

Minimum 10 cases:

Category	Include
☑ Normal walk tracking	Duration saved
☑ Manual entry success	Editable fields
☒ Permission denied	Clear alert + fallback
☒ Incomplete manual entry	Block save
☑ Persistence	History loads correctly
☒ Future timestamps	Block + inline message

Bonus: behavior when user minimizes the app ⏱

PART 5 — UAT WALKTHROUGH SCREENS

Export: 05_UAT_LogWalk.pdf

Demonstrate:

- ✔ Initiate walk
- ✔ Track & stop
- ✔ Save entry
- ✔ Verify in history

Stakeholder comments should include:

- Ease of starting a walk
- Whether tracking UI is intuitive
- Suggestion (e.g., distance indicator future enhancement)

PART 6 — SPRINT REVIEW MINI-SCRIPT

Export: 06_SprintReview_LogWalk.md

60–90 seconds:

- Value to users: "Helps pet owners monitor their dog's exercise & well-being."
- Live demo path: Start → End → History
- What improved from Sprint 2: Better validation + clearer navigation
- What's next: Medication tracking and real-time app health metrics

PART 7 — PORTFOLIO README UPDATE

Export: 07_Portfolio_README_Sprint3.md

Add:

- Screenshot of tracking state
- Callout of permission handling
- Note how behavior changes offline or minimized

Demonstrate growing tactical skill with each Sprint.

CHAPTER SUMMARY

In **Sprint 3**, Team Vertex successfully designed and delivered the Walk Log feature, bringing real-time tracking and reminders to the PetTracker app. This chapter showed the team diving deep into the **Design Phase** of the SDLC, where thorough planning and cross-functional collaboration were key. They turned an ambitious idea (a background-enabled timer with GPS) into reality by refining clear user stories, nailing down acceptance criteria, and iterating on thoughtful UI/UX designs. The development process wasn't without challenges: the team faced platform quirks and a serious bug, but their Agile practices (daily stand-ups, pair programming, and rapid testing cycles) allowed them to adapt and overcome each obstacle. Ultimately, Sprint 3 proved that Team Vertex could handle complex, time-sensitive features under pressure. They emerged not only with a valuable new app capability (scheduling walks and sending gentle reminders during a session), but also with stronger team discipline and new additions to their project toolkit (like the permission-handling checklist and reusable test cases). The Sprint's end saw the team documenting their learnings and improvements, ensuring that hard-earned knowledge is carried forward.

WHAT'S NEXT?

Sprint 3's success and lessons set the stage for an even more exciting Sprint 4. Having conquered real-time tracking, Team Vertex now turns to the next big user need: **Medication Tracking & Logging**. In the upcoming Sprint, the team will tackle features that allow pet owners to schedule and log medication doses. How will the team apply their improved processes to a new domain of the app? What new challenges will arise as they implement scheduling logic and notification systems? The next chapter will follow Team Vertex as they carry their momentum into Sprint 4, designing and delivering yet another crucial feature for PetTracker. Stay tuned: the journey continues with new goals, new obstacles, and the same tight-knit team determined to keep improving!

SPRINT 4

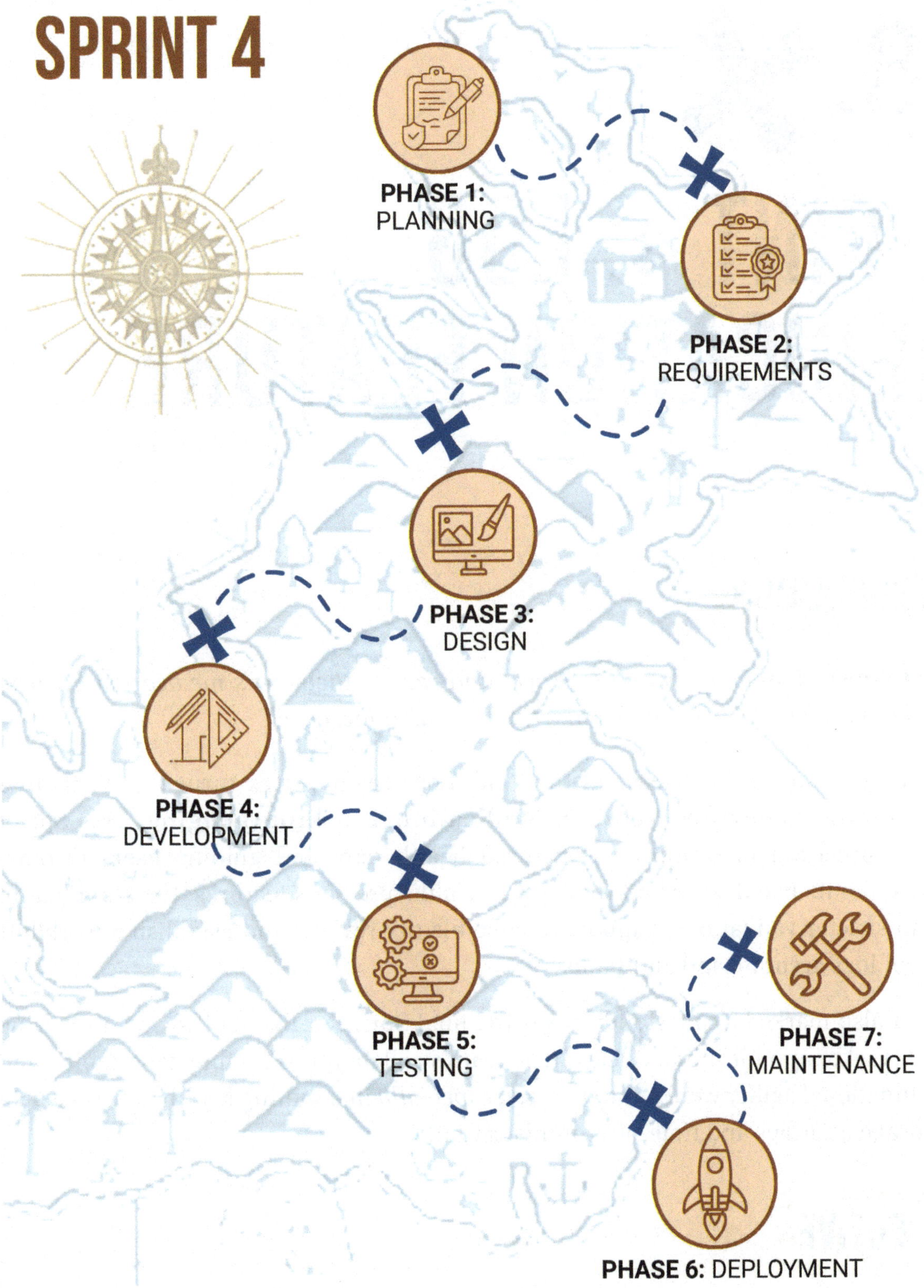
PHASE 1:
PLANNING
PHASE 2:
REQUIREMENTS
PHASE 3:
DESIGN
PHASE 4:
DEVELOPMENT
PHASE 5:
TESTING
PHASE 6: DEPLOYMENT
PHASE 7:
MAINTENANCE

18. SPRINT 4: LOG MEDICATION FEATURE

INTRODUCTION

Sprint Goal: *Enable users to log, track, and review their pet's medication history, ensuring accurate records for dosage, timing, and adherence."*

Following the successful release of the Walk Log, Team Vertex turned its focus to a feature central to proactive pet care: **Medication Tracking/Logging**. This Sprint required precision in capturing structured health data and guiding users through sensitive inputs like dosage and time given. Unlike previous features, the team had to build in stricter validations, support reliable history review, and establish a workflow that kept logs accurate and auditable.

Sprint 4 also marked a milestone: it was the final Sprint of the MVP scope, preparing the PetTracker app for initial release. The team's collaboration, refinement habits, and cross-functional agility were put to the test, *especially* in handling health-sensitive data like dosage accuracy and time-of-administration.

OBJECTIVE

In **Sprint 4**, Team Vertex dives deep into the testing phase of the SDLC. This chapter emphasizes quality assurance practices in Agile, from defining clear acceptance criteria and rigorous test cases to catching critical bugs early and refining processes through retrospectives. By focusing on building the Medication tracking or logging feature, the team learns how thorough testing and cross-functional collaboration ensure a reliable, user-trusted product increment. For Team Vertex, this Sprint highlights how a strong **Definition of Done** and proactive QA involvement can drive a high-quality release, ensuring a trustworthy health record for vet visits and long-term care.

FEATURE IN FOCUS: MEDICATION TRACKING/LOGGING

The Medication Tracking/Logging feature allows pet owners to record each dose of medication administered to their pets, track upcoming doses, and review past medication history.

Key focus areas:

- **Medication Entry:** Users can select a medication from a pre-set list or add a custom medication name.
- **Dosage & Frequency:** Fields for dosage amount, unit (e.g., mg, ml), and frequency (once, twice daily, etc.).
- **Time & Date Logging:** Ability to log when each dose was given.
- **Missed Dose Recording:** Option to log missed doses for accurate history.
- **Historical View:** Users can view a history log per pet, showing past and upcoming doses in chronological order.
- **Data Reliability:** Stored data is tied to each pet profile, ensuring accurate tracking per pet.

TEAM VERTEX IN ACTION

Planning the Sprint: JAD Session. To kick off Sprint 4, Maria (Product Owner) led a session with the team. With the MVP deadline in view, this final feature demanded careful attention to both usability and reliability. She framed the discussion with a key user story need:

> "AS A PET OWNER, I WANT TO LOG MY PET'S MEDICATIONS: NAME, DOSAGE, AND TIME GIVEN SO I

CAN TRACK ADHERENCE AND SHARE AN ACCURATE HISTORY WITH MY VET."

From this, the team aligned on a singular Sprint objective: build a robust, user-friendly scheduling system that supports the following scope:

- **Required inputs:** Medication name, dosage (value + unit), date & time given
- **Optional input:** Notes for additional context
- **Validation:** No required field blank; dosage must be valid; time given cannot be in the future
- **Cross-platform consistency:** Provide a consistent experience across iOS and Android apps

They agreed that this feature would be Sprint 4's sole deliverable (estimated at 8 story points). By narrowing focus, the team could maintain quality and give the scheduling feature the attention it deserved.

Workflow Considerations. During this planning session, Team Vertex reviewed how adding, editing, and viewing medication logs should work; clarified unit handling (mg, mL, tablet count), preventing duplicate entries within short time windows, and ensuring history is per-pet. Jenna (Designer) and Malik (QA Engineer) noted a few workflow gaps and edge cases to refine during the design and testing phases.

MEDICATION LOG WORKFLOW DIAGRAM

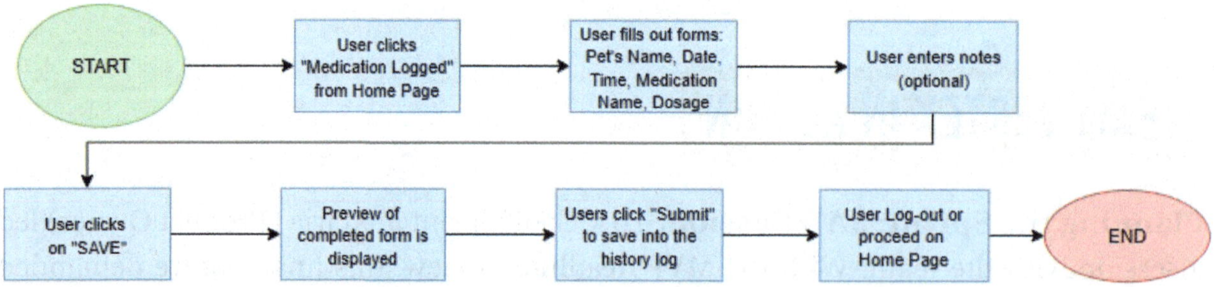

Commitment. At the close of the kickoff, Team Vertex committed fully to the plan. More than just closing out another Sprint, this feature would round out the PetTracker MVP with care, clarity, and confidence. The team was energized to build a feature that would help users manage medication logging reliably.

Requirements: Breaking Down the Feature

Every feature in Scrum is a landscape to be mapped. By breaking down *Medication Tracking & Logging* into clear user stories, the team can navigate the complexity of medication management one step at a time. Each story becomes a waypoint on the journey, helping the Product Owner and Development Team focus on delivering functionality that supports accuracy, compliance, and user confidence.

FEATURES TO STORIES BREAKDOWN

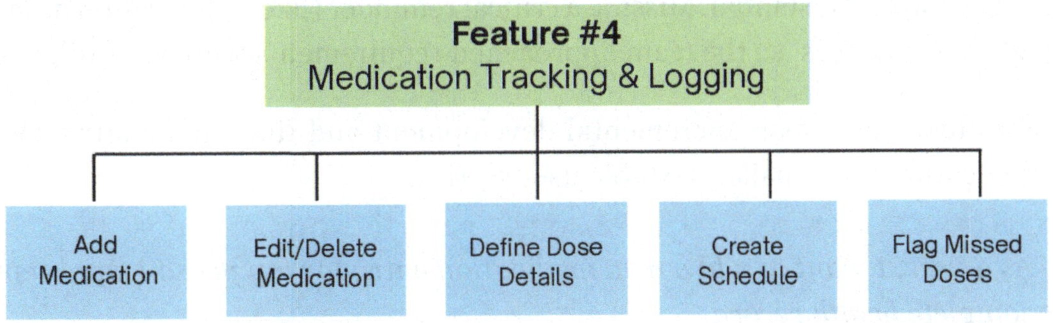

This breakdown illustrates how the high-level feature **Medication Tracking & Logging** is decomposed into discrete user stories to enable iterative delivery and refinement. Each story represents a specific slice of functionality:

- [x] adding medication
- [x] editing or deleting records
- [x] defining dosage details
- [x] creating schedules
- [x] flagging missed doses

that collectively achieve the feature's goal.

By organizing the work at this detailed level, the Product Owner and Development Team can prioritize stories based on their value, refine acceptance criteria, and deliver testable increments that align with stakeholder needs. This approach ensures transparency, adaptability, and a clear pathway from the feature vision to potentially shippable product increments that collectively fulfill the feature's objective. By structuring the work at this level of granularity, the Product Owner and Development Team can prioritize stories based on value, refine acceptance criteria, and deliver testable increments that are aligned with stakeholder needs. This approach ensures transparency, adaptability, and a clear path from feature vision to potentially shippable product increments.

With Medication Tracking/Logging as the Sprint's goal, Maria refined the core feature into a primary user story:

> *Feature:* As a pet parent, I want to log medications so I can track treatment schedules and adherence.

Though the story sounded straightforward, its implications for timely care, data reliability, and user trust were substantial. Missing a critical reminder (like a medication dose) could have real consequences, so the team approached requirements with extra diligence.

User Stories: To support incremental development and thorough testing, the team broke the feature into smaller, testable user stories:

- *As a user, I want to add a new medication entry for my pet so I can keep a complete health record.*
- *As a user, I want to log each administered dose with date and time so I can track adherence.*
- *As a user, I want to mark a dose as missed so I can maintain accurate records.*
- *As a user, I want to see my pet's medication history so I can provide accurate information to the vet.*
- *As a user, I want dosage fields validated so I cannot save without required medication details.*

Each story was documented in Jira and estimated during Sprint planning. This breakdown allowed design, development, and QA to work in parallel with clear boundaries for each functionality.

Acceptance Criteria: To meet the Definition of Done, the team established clear conditions for each story and the feature as a whole. Key acceptance criteria included:

- **Required Fields:** Medication name (must be specified for a dose log entry to be set)
- **Optional Field:** Notes (additional details or instructions, not mandatory)

- **Validation Rules:** Required fields cannot be blank; dosage must be a valid positive number; time given cannot be in the future.
- **Logging Logic:** Each dose log entry must be tied to the selected pet's profile. Entries tied to selected pet; timestamps UTC-normalized. Medication log entry will be editable or repeatable in a future release (beyond MVP).

During backlog refinement, the team discussed open questions to clarify requirements:

- **Malik** asked whether duplicate medication entries or multiple dose logs at the same exact time should be prevented, to avoid confusing duplicate notifications.
- **Jenna** needed design direction on dosage control: numeric input + unit selector vs free text.
- **Alex** proposed using a unified time standard (UTC) behind the scenes for storing all dose log times, to avoid any future issues with time zones or daylight savings when the data is synced or if users travel.

Maria considered these points and confirmed the approach for the MVP:

Overlapping events **would be allowed** for now; the system would not block users from setting multiple reminders at the same time, leaving it to user judgment.

The MVP would use simple, native **date and time pickers** for scheduling inputs (no complex calendar grid interface yet), to keep things straightforward.

UTC standardization for stored timestamps was approved for backend consistency, ensuring reminders remain accurate regardless of device locale.

By the end of the requirements phase, Team Vertex had:

- A clear backlog of refined, testable user stories.
- Defined acceptance criteria and validation logic for each scenario.
- Early clarity on design and data concerns (like time zones and UI behavior).
- Reduced ambiguity going into the design and development phases.

This upfront clarity gave Sprint 4 the structure needed to meet both the MVP's goals and long-term reliability standards.

Design: UI/UX with an Eye on Consistency

At the start of Sprint 4, Team Vertex convened for Sprint planning with a clear, tightly scoped objective: deliver a **medication tracking and logging interface** that was intuitive, visually consistent with the PetTracker design system, and compliant with all acceptance criteria.

Jenna opened the session by walking the team through her initial sketches and user flow assumptions for logging doses and tracking medication history. This ensured alignment before any development began.

MEDICATION LOG PROTOTYPE

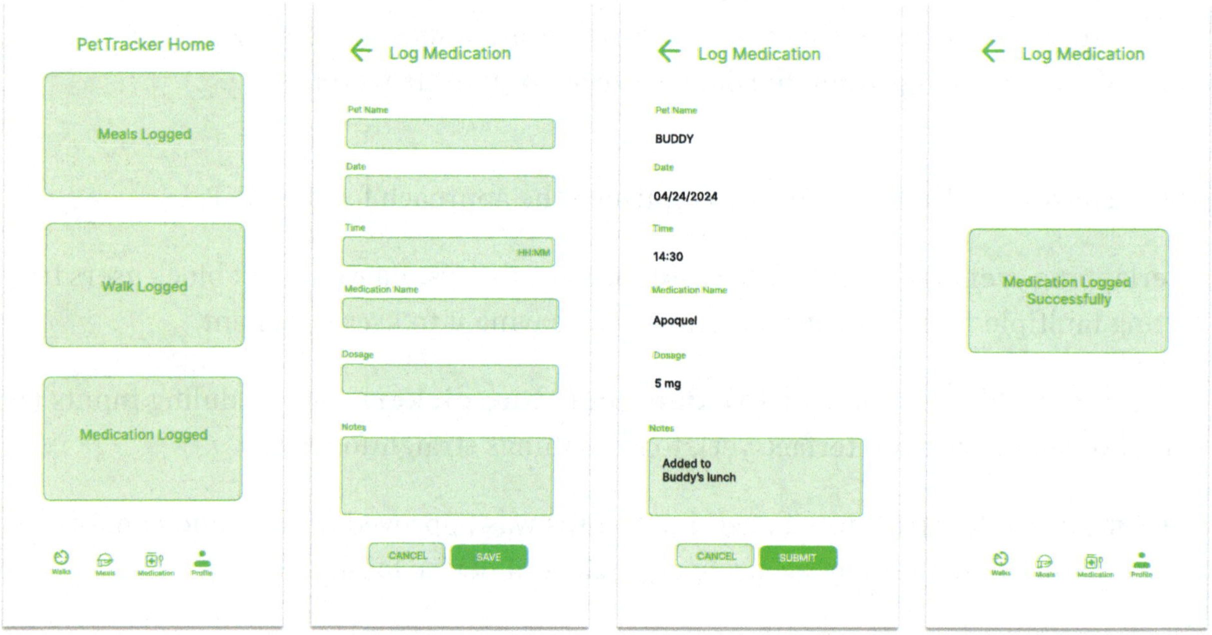

UI Design: Familiarity & Focus

Jenna designed a clean, mobile-friendly logging form, deliberately building on visual patterns established in earlier Sprints, particularly the **Feeding Log** from Sprint 2. This created a sense of familiarity for users and reduced cognitive load.

The medication log form included the following core fields and controls:

- **Medication Name**: Free-text input for identifying the drug or treatment (e.g., "Flea Pill" or "Antibiotic: Amoxicillin").
- **Dosage**: Numeric field with unit selector (e.g., mg, ml, tablet) for accuracy.
- **Date Administered**: Date picker defaulting to today's date, with restrictions against selecting future dates.
- **Time Administered**: Time picker defaulting to the current time to streamline entry.
- **Notes**: Optional text area for instructions or observations (e.g., "Give with food," "Vomited after dose").

To support accurate and efficient logging, Jenna introduced several usability enhancements:

- **Clear labels and placeholder text** for each field to guide data entry.
- **Visual indicators** (asterisks or red outlines) marking required fields (Medication Name, Dosage, Date/Time).
- **Native pickers and dropdowns** to prevent invalid entries (e.g., restricting dosage input to positive numbers).
- **Defaulting to the current date/time** for quick logging immediately after administration.

Sprint Readiness Review

Before moving forward, the team conducted a Sprint readiness check:

- Malik flagged an edge case: what if a user attempted to log a future date by mistake? They agreed validation rules would block future timestamps.
- Alex confirmed that the new MedicationLog component could reuse existing form modules from previous Sprints, reducing build time.
- Maria reviewed Jenna's mockups against the acceptance criteria, ensuring error states and edge cases were visually represented.
- Once all items were addressed, Jenna attached the approved designs to the Jira story and marked it "Design Complete."

The Definition of Ready checklist was fully met:

- UI mockups attached and approved.
- All validation rules documented.
- Data model for medication history aligned with backend storage plans.
- QA test scenarios pre-drafted for each acceptance criterion.

With planning complete and a unified visual approach, Team Vertex moved confidently into the build phase.

Development: Building & Internal Testing

With the design finalized and user stories refined, Sprint 4's development phase began with a focus on **data accuracy, reliability, and clear error handling**: essential for health-related features.

Core Development Tasks

Alex (Developer)

- Built the **MedicationLog** component using the team's existing input, picker, and validation modules for consistency.
- Integrated date and time pickers with defaults and validation to prevent future dates or incomplete selections.
- Implemented logic to persist each medication log to the backend under the correct pet profile, ensuring historical accuracy.
- Created backend hooks to support filtering logs by medication type or date range for future analytics features.

Malik (QA)

- Developed front-end validation to enforce required fields and prevent saving incomplete or invalid logs.
- Identified and helped resolve a bug where leaving the dosage unit blank bypassed validation.
- Verified that each saved log appeared immediately in the pet's medication

history, with a success message or error prompt when needed.

XP PRACTICE #4 – TEST-DRIVEN DEVELOPMENT: BUILDING FOR CONFIDENCE

Sprint 4 introduced a new development rhythm: Write the test first. Prove it fails. Then make it pass. Instead of writing the Medication Log logic blindly and hoping QA would catch edge cases, Alex flipped the script:

1. **Wrote failing unit tests for missing dosage**, future date input, incomplete time selection, and negative values.
2. Then wrote just enough code to pass each test.
3. Only after green ✅, the feature moved forward.

This was Team Vertex's introduction to Test-Driven Development (TDD) – their fourth XP discipline.

What changed?

- No more "write everything, test later"
- Clear confidence the validation logic couldn't regress
- Every bug became a new test

Malik added even more test scenarios once Alex's logic passed:

- Optional Notes field saves only when required fields are valid
- Invalid date → form blocks submission
- Correct save triggers success toast and updates history view

Maria called it out in Review: "This feels rock solid. Like...you can't break it even if you try." That's the power of TDD.

TDD in action: Build with confidence

Before (Failing Test)

javascript Copy code

```javascript
// should reject future date
expect(isValidDate(inputDate)).toBe(false);
```

After (Passing Implementation)

javascript Copy code

```javascript
function isValidDate(inputDate) {
  return inputDate < Date.now();
}
```

Daily Scrum Highlights

Day 2: Standup Summary:

Maya (Scrum Master): "Alex, any blockers?"

Alex (Developer): "Time picker is now respecting device locale. Working on ensuring backend entries are sorted correctly in history view."

Malik (QA): "Validation is solid, but the Notes field isn't saving if other fields fail validation, which is expected, but I'll recheck after Alex finishes sorting logic."

Maria (PO): "I'll run a user-flow test later today. Also spotted a missing label on the Dosage Unit dropdown. Jenna?"

Jenna (Designer): "Got it. I will update and push the fix before lunch."

OUTCOMES BY END OF DEVELOPMENT PHASE

By Sprint's end, the team delivered:

- A **fully functional Medication Tracking feature** across iOS and Android.
- Accurate **logging and storage** of medication name, dosage, date, time, and notes under the correct pet profile.
- Strict validation rules to avoid future dates, negative dosages, or missing required fields.
- Clean error messaging guiding users to correct issues before saving.
- A modular codebase ready for future enhancements, such as refill tracking or medication reminders.

XP PRACTICE #5 – CONTINUOUS INTEGRATION: QUALITY WITH EVERY COMMIT

By Sprint 4, Team Vertex had multiple features live: authentication, feeding logs, GPS tracking, and now medication tracking. It was time to stop relying on hope and manual checks alone. They implemented their fifth XP practice:

Continuous Integration (CI): every commit builds, tests, and alerts.

What CI meant for the team:

- GitHub Actions ran linting, unit tests, and build verification on every pull request
- If anything failed, the PR was blocked
- Only green builds made it to develop
- CI also ran on merge to catch integration errors across modules

Malik celebrated it first:

"I can trust that if the pipeline is green, nothing major broke in auth, logs, or history view."

Alex noticed something else:

"I'm fixing bugs before you even file them."

CI gave the team confidence, speed, and a shared heartbeat of quality.

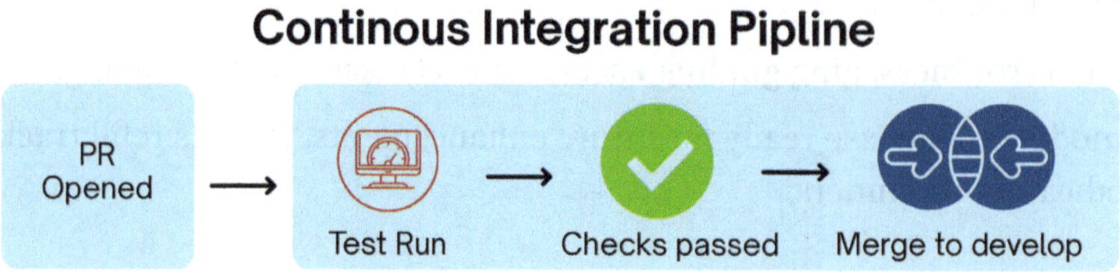

This diagram illustrates Team Vertex's Continuous Integration (CI) pipeline. When a developer opens a pull request, automated tests are immediately triggered. If all checks pass—linting, unit tests, and build verification—the code is cleared to merge into the develop branch. This ensures every new feature integrates safely, maintains quality, and keeps the team moving fast without breaking shared progress.

TESTING: VALIDATING WITH PRECISION

As the **Medication Tracking** component reached code completion, Malik led the testing phase with a laser focus on accuracy and edge cases. Given that this feature handled **health-critical data** like logging what medication was given, when, and in what dosage, there was zero tolerance for errors that could compromise user trust or lead to missed or incorrect dosing. Testing was treated with the same weight as development itself.

MEDICATION LOG APPLICATION BUG

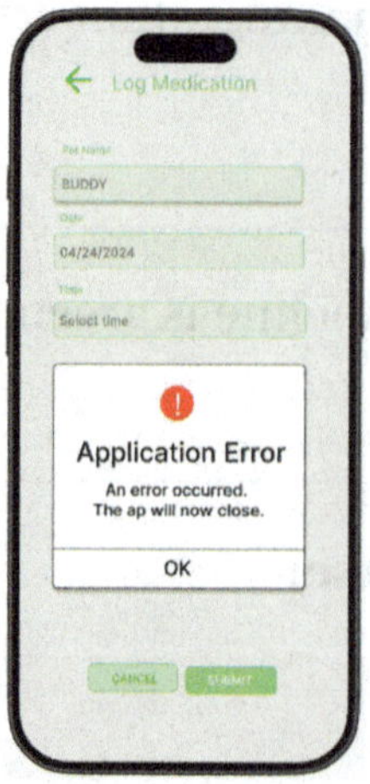

Critical Bug Discovered

Early in the testing cycle, Malik found a serious issue. While testing form submissions, he discovered that if a user selected a **date** but skipped the **time administered** field, tapping "Save" caused the app to crash.

Reproducing the crash on multiple devices, Malik identified the cause:

1. The logging form's logic assumed both date and time would always be filled in.
2. The backend expected a complete timestamp for every medication entry.

Without a time value, the submission pipeline broke, and the app failed to handle the error gracefully.

Alex's Fix:

- Added a **default time** (current device time) if the user left the time field blank, ensuring no null timestamps were ever sent.
- Updated validation logic to prompt the user if either the date or time was missing, preventing incomplete submissions.

This fix not only resolved the crash but also reinforced the rule that a **medication log cannot be saved without a complete administration timestamp**.

Expanded Test Coverage

With the bug resolved, Malik broadened the test suite to address related scenarios, including:

- **Blank field submissions**: Ensuring missing required fields (Medication Name, Dosage, Date/Time) triggered user-facing errors without crashes.
- **Invalid inputs**: Simulating nonsensical dosage values (negative numbers, unsupported units) and blocked them via validation.
- **Past/future date handling**: Confirming that future administration times were disallowed, while past logs could be entered for record-keeping.
- **Time zone consistency**: Verifying that an "8:00 AM" entry stored in UTC still displayed and sorted correctly in the user's local time.

- **Successful logging flows**: Ensuring a properly completed form stored the medication log under the correct pet profile and immediately updated the history view with a success confirmation.

Malik combined automated unit tests with **manual device testing** on both older and newer phones across iOS and Android. He confirmed that error messages displayed clearly, entries persisted correctly, and history listings remained consistent.

UI TESTING: IT'S NOT JUST ABOUT LOOKS—IT'S ABOUT TRUST

While Sprint 3 drilled into functional stability, Sprint 4 brought the user interface into sharp focus. The Medication Log introduced form fields, time pickers, error messages, and visual confirmations—all of which needed to work not only technically but intuitively.

Malik shifted QA toward UI testing to ensure:

- Consistency across devices (e.g., time picker layout on iPhone SE vs Pixel)
- Clear visibility of error messages when required fields were left blank
- Form alignment and spacing remained intact regardless of screen size or orientation
- Touch targets were sized correctly for quick entry, even in one-handed use
- Confirmation messages and history updates were immediate and visually reassuring

Malik also verified color contrast, button spacing, and tap responsiveness to catch any visual friction points—especially for time-sensitive, high-stakes inputs like medication tracking. This attention to UI precision helped reinforce confidence in the app's reliability and ease of use.

Quality Lesson Learned

This incident underscored an important principle: **Even required fields need intelligent defaults and fail-safes.** Missing or malformed data should never break functionality.

The team is committed to adding similar "graceful fallback" rules to all health-sensitive features in the app going forward. These considerations would now be part of acceptance criteria and test planning from the start.

UAT DEPLOYMENT: READY FOR STAKEHOLDER VALIDATION

At Sprint 4's close, the **Medication Tracking** feature was deployed to the shared UAT (User Acceptance Testing) environment: a staging space where internal stakeholders could interact with the feature in a production-like setup.

Accompanying the UAT release:

- **Alex** deployed the build after passing all CI pipeline checks.
- **Jenna** reviewed UI layouts for various screen sizes, ensuring dosage fields, date pickers, and history lists displayed cleanly on both small phones and tablets.
- **Malik** conducted final end-to-end submission tests, confirming logs were stored under the correct pet, timestamps were accurate, and data persisted without corruption.
- **Maya** drafted release notes summarizing the feature, marking the Jira ticket as "Ready for Review" and notifying stakeholders.

SPRINT REVIEW: LIVE DEMO

In the Sprint Review meeting, stakeholders tested the Medication Tracking feature in real time:

- Created sample logs such as "Flea Pill: 5mg" and "Antibiotic: 1 tablet," adding notes to some and leaving notes blank for others.
- Malik monitored backend logs, confirming each entry was saved under the correct pet with accurate date/time stamps.
- He demonstrated the history view updating instantly after each log, using a 1-minute-backdated entry to show how recently administered doses appeared.
- **Maria** checked every acceptance criterion: required field validation, proper sorting of history entries, and correct association of logs to the pet profile. Satisfied, she confirmed the feature was MVP-ready.

Stakeholder Feedback:

Positive and enthusiastic. They highlighted how vital this would be for busy pet owners managing multiple medications. One forward-looking request was made: to add **recurring dose templates** for pets on long-term medication. The team added "recurring medication logging" to the backlog for future development.

With Sprint 4's feature complete, validated, and reviewed, **Medication Tracking (Feature 4)** stood as a fully tested, production-ready MVP component ready for official release and trusted by pet owners to keep accurate health records.

AGILE IN MOTION

Sprint 4 showcased several Agile and SDLC best practices in action, especially around quality assurance and team collaboration:

- **Collaborative Planning:** The team began with a Joint Application Development session involving all roles. This upfront collaboration ensured that potential pitfalls (like recurring events or UI concerns) were identified early and that everyone shared a clear vision of the feature.
- **Focused Scope & Estimation:** Team Vertex committed to a single, well-defined Sprint goal and user story (estimated at 8 points). Keeping the scope tight helped maintain quality. The team's velocity and past experience informed this estimate, ensuring the work was achievable within the Sprint.
- **Incremental Story Slicing:** The feature was broken into bite-sized user stories with clear acceptance criteria. This made development and testing more manageable and allowed the team to deliver value in increments, verifying each piece (from input validation to notification handling) independently.
- **Definition of Ready & Done:** The team rigorously applied their Definition of Ready before development (for example, confirming designs, data models, and test scenarios were prepared). Similarly, the Definition of Done required that all acceptance criteria be met and all tests (including regression tests) pass before calling the feature complete. This disciplined use of DoR/DoD ensured no aspect of quality was overlooked in the rush to finish the Sprint.
- **Continuous Testing and QA Integration:** Quality assurance wasn't an afterthought, as Malik was involved from the start, contributing to requirements and design discussions. He wrote test cases in parallel with development and

performed extensive exploratory testing. When a critical bug appeared, the team treated it as a learning opportunity, quickly fixing it and expanding test coverage. This test-driven mindset led to a more robust feature.

- **Cross-Platform & Device Testing:** The team recognized the importance of the app working on all devices. They included cross-platform checks (iOS and Android) and even different device types in their testing plan. This proactive approach caught issues that might only appear on certain screens or OS versions, improving overall product quality.
- **UAT & Fast Feedback:** Instead of going straight to production, the team used a UAT environment to gather stakeholder feedback and do a final validation in a safe space. This practice reflects Agile's emphasis on getting fast feedback and involving stakeholders throughout. It helped catch any last-minute adjustments and built stakeholder confidence prior to launch.

Throughout the Sprint, **continuous improvement** was evident. The team held daily stand-ups to adapt to micro-blockers and ended with a thorough retrospective to improve their process, embodying the spirit of Agile in motion.

TEAM REFLECTION

After wrapping up Sprint 4 (and with the entire MVP nearing launch), Team Vertex came together for a retrospective that was equal parts celebration and constructive analysis. The mood was proud but never complacent: the team wanted to glean every lesson from their journey so far.

Successes: They noted several high points. Every Sprint, including this one, resulted in a **demo-ready increment**, meaning they always had working software to show. This consistent delivery built trust with stakeholders and momentum within the team.

Additionally, thanks to their diligent testing in Sprint 4, the app was remarkably stable. In fact, no major crashes were reported in the first weeks of testing the full MVP. Morale was high, knowing that their hard work on quality had paid off in a smooth, reliable product.

Friction & Lessons: The team openly discussed what could be improved. A few early bugs across the project were traced back to incomplete or vague acceptance criteria, highlighting the need for even clearer specs during refinement.

They also realized that **cross-platform UI issues** had slipped through (for example, a minor layout quirk on a smaller iPhone screen), a reminder to broaden their device testing matrix.

Another point was that some edge cases hadn't been fully defined upfront in certain user stories, causing last-minute scrambles. These honest acknowledgments weren't about blame, but about learning as a team.

Adaptations: To address these points, Team Vertex decided on some concrete action items. One was introducing an **Acceptance Criteria review checklist** in their refinement process. Essentially, a tool to ensure every story's criteria cover all input fields, errors, and edge cases (they actually started using such a checklist during Sprint 4's development when they realized its value).

Another improvement was to diversify their test device coverage: Malik suggested maintaining a rotating list of devices (including older phones and low-memory models) to test on, so issues aren't only caught on the most modern hardware. The team also talked about formalizing Definition of Ready even further, to include explicit consideration of edge cases.

In fact, the retrospective closed with each team member making a personal commitment for the next development cycle:

Maya (Scrum Master): "I'll refine our story readiness definition." (She would ensure that before a story enters a Sprint, it meets an even more robust checklist of readiness.)

Maria (Product Owner): "I'll partner earlier with QA during story grooming." (She plans to involve Malik when defining acceptance criteria, to inject a testing perspective from the start.)

Alex (Developer): "I'll build error logging into all new features by default." (He wants to make debugging easier by ensuring the app logs any failure points.)

Jenna (Designer): "I'll finalize the mobile UI kit for faster mockups." (She intends to firm up a reusable design component library to speed up design and maintain consistency.)

Malik (QA Engineer): "I'll double down on validation testing and edge case simulation." (He commits to push even more on negative testing and weird scenarios, to catch problems before they hit users.)

These pledges reflected the team's growth mindset. Even after a successful Sprint and MVP completion, everyone was eager to improve their craft. The honest dialogue and mutual support in this retrospective boosted team morale: they could see how far they'd come as a cohesive unit, and they were excited about where they could go next.

Exercises & Activities

Sprint 4 Exercise Pack

Medication Log + Reminder Compliance

Goal: Help users track and prove medication adherence for their pets.

Students now think like Product Builders — focusing on usability, risk reduction, and health compliance.

Deliverable Checklist: Export each file with the exact filenames listed below

Part 1 — Activity / Workflow Diagram

Feature Flow Requirements:

- Enter medication entry (name, dose, date/time)
- Toggle "Taken now" vs "Scheduled reminder"
- If scheduled → reminder created → confirmation state

Error handling:

- Missing required fields
- Invalid dates/times
- Save + history navigation

Export: 01_Activity_MedicationLog.pdf

Part 2 — User Stories + Acceptance Criteria

Minimum 4 stories (Given / When / Then format)

Required coverage:

- Log medication dose ✅
- Schedule reminder with date/time ⏰
- Track "Taken vs Missed" 🔁
- Show medication history 📈

Export: 02_Stories_AC_MedicationLog.pdf

Part 3 — Low-Fidelity Wireframes

Screens required:

1. Add Medication Entry
2. Reminder Scheduling
3. Medication History View

Annotations:

- Required fields labeled
- Error states visible
- Reminder success banner

Export: 03_Wireframes_MedicationLog.pdf

Part 4 — Test Case Matrix

Minimum 10 test cases covering:

✅ **Valid entry**

- ❌ Missing dose or time
- ⏰ Reminder incorrectly set in the past
- 🔁 Missed reminder → status update
- 📱 Persistence after restart
- 🚫 Cancel / discard a draft entry

Export: 04_TestCases_MedLog.csv (or PDF)

Part 5 — UAT Walkthrough

Show the core experience like a real user:

- ✔ Add medication entry
- ✔ Schedule reminder
- ✔ See medication appear in History
- ✔ Screenshot proof

Include stakeholder comments such as: "Reminder UX is clear — next step: snooze option?"

Export: 05_UAT_MedLog.pdf

Part 6 — Sprint Review Script

Short, confident, and business-value language: What shipped, why it improves pet health outcomes, and mew risks → recommendations.

Export: 06_SprintReview_MedLog.md

Part 7 — Portfolio README Update

Add Sprint 4 artifacts to the running summary. Include screenshots, lessons learned, and pet health compliance value proposition

Export: 07_Portfolio_README.md (version-incremented)

Bonus Stretch Challenges

Optional but signals higher maturity

- Include medication categories (pill, liquid, injection)
- Include voice input idea as backlog enhancement
- Add test automation plan (placeholder for CI integration)

Export: 08_Stretch_Sprint4.pdf (only if attempted)

CHAPTER SUMMARY

Sprint 4 delivered a critical new capability: **Medication Tracking & Logging**. By the end of this Sprint, Team Vertex had a fully tested feature that lets pet owners plan care events and receive timely alerts, adding a proactive dimension to the PetTracker app. Importantly, this Sprint wasn't just about the feature itself, but about cementing quality practices. The team applied rigorous testing procedures, caught and fixed a major bug before release, and ensured the feature met a high standard of reliability.

This Sprint also rounded out the Epic #1: PetTracker MVP feature set. All planned core features were complete and reviewed, giving the team confidence that the app was ready for real users. The collaborative effort in Sprint 4 reinforced the team's agile mindset: open communication, quick iteration on issues, and continuous improvement. In terms of outcomes, the scheduling feature was deemed MVP-ready (having passed UAT with stakeholder approval), the team's QA processes proved their worth (with a stable, crash-free build), and Team Vertex grew even more cohesive through the lessons learned.

WHAT'S NEXT?

With the testing phase behind them and the MVP features built, Team Vertex faces the ultimate challenge: **launching PetTracker to the world**. In the next chapter, the focus shifts to releasing the app into production and monitoring it in the wild. How will the team handle real user feedback and the pressures of a live product? The upcoming Sprint will explore the deployment process, live-site monitoring, and the first post-launch enhancements as the PetTracker journey continues. Are they truly ready for launch day and beyond? The next chapter will tell.

19. PRODUCTION DEPLOYMENT/MVP RELEASE

With Sprint 4 closing soon, Team Vertex is ready for its biggest milestone yet: releasing the PetTracker MVP to production. After the Sprint Review and retrospective, excitement is high, but Maya (Scrum Master) reminds everyone that a smooth launch takes careful planning.

She calls a release-readiness meeting, whiteboard markers in hand.

Maya (Scrum Master): "Before we say this feature is live, let's run through our pre-release checklist: nothing should surprise us on launch day." The team nods.

They know what it's like to fly blind and want this go-live to be calm and controlled. Maya reads off key items: change log complete, risks assessed, rollback plan ready, stakeholders alerted, tests validated, monitoring in place.

Changelog & Versioning: All changes in the release are documented (in Jira, Git, or a CHANGELOG file) so everyone knows what's new. This becomes the source for release notes and troubleshooting later. **Alex (Developer) and you** double-check that all commit messages and tickets for this Sprint are closed and noted.

Risk Assessment & Rollback Planning: The team reviews potential risks (e.g., server load, data migrations, security) and how to handle them. Maya updates the risk log with any new items: what could go wrong and how to revert. **Malik (QA)** suggests a rollback test: *"If login breaks in prod, do we have a backup of the old release to switch back?"* Alex confirms they'll keep the previous Docker image handy.

Stakeholder Notifications: Maria (Product Owner) prepares communications. She writes a brief release note outlining the MVP features and known limitations. She schedules an email to stakeholders (marketing, customer support, executives) notifying them of the go-live plan and expected timing. **Jenna (UX)** volunteers to create a short help doc snippet for user-facing teams.

Test Validation: Malik (QA) ensures all automated tests have passed and that a final regression suite has run cleanly. The team performs one last smoke-test on their staging (UAT) environment, simulating critical user flows (sign-up, login, pet profile). Any issues found are fixed immediately. Malik signs off only when he's confident everything is solid.

Monitoring & Alerts Setup: Maya confirms that production monitoring is ready. Alex has configured logging and alerts (errors, performance). They ensure they have dashboards for key metrics (user sign-ups, API errors). Maya adds a monitoring rule: notify the team if error rates spike. Jenna checks that client-side analytics is collecting data on the new screens.

Together, these practices give Team Vertex confidence that they aren't forgetting anything obvious. Maya smiles at the filled-out checklist on the board and says, *"Good. We're as ready as we can be. Let's review who does what on deployment day."*

Team Roles in Release Readiness

Every role on Team Vertex has a part to play in the release. At the meeting, Maya asks each member to confirm their responsibilities. Teaming up for a quick round-robin: who will do what?

Product Owner (Maria): Defines the release scope and "go/no-go" criteria. Maria ensures the backlog accurately reflects what's being deployed (and what's deferred). She finalizes release notes and communicates with stakeholders (marketing, sales, customer support) about the launch plan and features. During the release, she's ready to answer questions from management or clients.

Scrum Master (Maya): Coordinates the entire deployment process. Maya makes sure the team follows the pre-production checklist, tracks tasks on the Scrum board, and removes any roadblocks. She schedules the release window and any necessary meetings (e.g., launch call). On deployment day, Maya facilitates communication between developers, testers, and any ops or support staff. Afterward, she leads the quick post-release micro-retrospective.

Developers (Alex & You): Prepare and execute the technical release steps. Alex and you (the Backend/API Developer) will merge the code to the release branch or main line, trigger the CI/CD pipeline, and oversee the production deployment. You ensure that any feature toggles are set correctly (new features off or on as planned). You monitor the deployment job and confirm the build succeeded. If a rollback is needed, Alex has the procedure ready to execute.

QA (Malik): Leads quality checks. Malik runs smoke tests on staging right before deployment and on production immediately after. He verifies critical user flows (sign-up, login, profile creation) still work end-to-end. Malik watches error logs during and after the release. He's the go-to for any bug that slips through, ready to triage issues.

UX/Design (Jenna): Finalizes user-facing polish. Jenna double-checks that the user interface looks correct in production: no layout issues, missing images, or text errors. She updates any contextual help or tutorials with the new features. Jenna also collaborates with Malik to ensure the user experience is clear (for example, confirmation messages appear after saving a profile).

Support Team: (BrightByte's customer support and ops staff) The support group is briefed on the release. Maria or Jenna provides them with key talking points and troubleshooting tips for the new MVP. Support is ready to field user questions or incident reports. Meanwhile, an operations engineer (if available) stands by to handle any infrastructure issues (server restarts, load balancer checks, etc.).

By clearly defining these roles, Team Vertex ensures nothing falls through the cracks.

For example:

- Alex reminds: "After we merge, I'll deploy the code."
- Malik will run the first smoke test.
- Jenna will confirm the UI.
- Maria will hit send on the stakeholder email as soon as it's live.

Everyone knows their responsibility.

DEPLOYMENT DAY FLOW

The morning of the release, the team gathers (virtually or in person) ahead of the planned deployment time. The air is a bit tense, but Maya keeps spirits high.

Maya (Scrum Master): "Alright, everyone, deployment clock is ticking. Let's go through the steps to keep us on track."

1. **Staging Smoke Test:** Malik performs one more smoke test on the staging (UAT) environment. He walks through signing up, logging in, and creating a pet profile: the core flows. You, Alex, and Jenna watch. Everything passes. Malik gives a thumbs-up: *"Staging is green. All set."* This final check reassures the team that they're deploying solid code.
2. **Trigger Production Deployment:** Alex (Developer) merges the tested code into the production branch. He then clicks to run the CI/CD pipeline for production. Automated jobs compile, run tests again, and if all clear, deploy to the live servers. Jenna and you monitor the build logs with a bit of suspense. Within minutes, Alex reports, *"Pipeline succeeded: version 0.1 is live in production!"* Maya smiles: *it's good news.*
3. **Activate Features & Toggles:** The team uses feature toggles to control exposure. Alex ensures that any new features meant for the MVP are toggled "on" now that the deploy is done. For features not ready, the toggle stays "off," so incomplete code remains hidden. Malik immediately checks that enabling the toggles didn't break anything: he logs into production, creates a test pet, edits it, and all works. Jenna observes the UI elements appear as they should.
4. **Release Note Communication:** With the deployment successful, Maria sends the green-light email to stakeholders. Her subject line reads: "PetTracker MVP is Live!" Inside, she summarizes what's available (user registration, pet profiles) and what's still to come (multi-pet support, forgot-password, etc.). She thanks everyone for their patience and invites feedback. Simultaneously, the customer support team publishes a short bulletin on the internal portal explaining the release and giving tips for any known issues.
5. **Post-Release Micro-Retrospective:** Shortly after the deploy (or at day's end), Maya calls for a quick stand-up retrospective. The team sits down or jumps on a video call.

Maya: "Congratulations on a successful launch, folks. Before we break, let's take 5 minutes: what went well and what can we improve?"

Malik notes how smoothly the staging tests went thanks to prepping smoke-test scripts. Alex mentions that a last-minute pipeline config change was needed, but overall deployment was fast. Jenna appreciated having design final checks in place. They jot down one or two "start/stop/continue" bullets (for example, start log monitoring earlier, continue having these quick devops checks). The session is brief and upbeat: a celebratory high-five for the win and a couple of lessons for the next release.

The deployment day flows like a well-rehearsed plan. Clear roles, checklists, and open communication mean the team stays calm even when the stakes are high. As Maya said earlier: "A smooth launch feels easy when it's planned thoroughly."

Post-Deployment Follow-Up

With the MVP in production, Team Vertex shifts gears from "deploy" to "monitor and document." The day after release, the team reconvenes for a wrap-up check. Maya leads a short session to hand off monitoring and tidy up.

Monitoring & Support Handoff: Malik monitors the production logs and metrics for a full 24 hours, watching for errors or abnormal behavior. You and Alex keep an eye on server performance (CPU, memory, response times) to catch any scaling issues. Maya confirms that alerts (e.g., via Slack or email) will notify the team if any critical failures occur. Meanwhile, the support team has started triaging user feedback. Any bug reports or urgent questions from real users are logged and prioritized. If something critical arises, Maya will spring into action to coordinate a patch.

Backlog and Changelog Alignment: Maria updates the product backlog and Sprint board to reflect what was delivered. All user stories deployed are marked "Done" and moved off the Sprint board. She also updates the project's changelog document with the final version number and notes. If any features were half-finished or deferred (for example, multi-pet support was left for later), those are clearly flagged as upcoming backlog items. This alignment ensures the team's records match reality.

Knowledge Capture: Finally, the team captures what they learned from the release. Malik adds any new test cases discovered during deployment to their QA checklist. Jenna updates the design system documentation if UI tweaks were made. Alex and you write a short postmortem entry: what took longer than expected, what went smoother than anticipated, and any surprising issues encountered. Maya writes a brief summary in Confluence or the team wiki titled *"Production Release 0.1: Lessons"* so new team members can learn. This documentation, from troubleshooting tips to timing notes, will make their next release even easier.

These follow-up tasks tie a bow on the deployment. Team Vertex recognizes that delivery isn't over once the code is live; continuous monitoring, clear documentation, and backlog updates are all part of finishing the story.

KEY LESSONS

Team Vertex's first production launch reinforced some timeless Agile principles and best practices. As they reflect on the experience, they highlight:

"Done" ≠ Deployed: Just because the team checked off all development tasks doesn't mean they delivered value. Team Vertex learned that a feature is only truly done when it's live for users. In other words, shipping the code (and confirming it works in production) is as important as writing it. They reminded each other that no one benefits until the deployment is complete.

Calm launches are planned: A successful launch felt calm, but it was no accident. Maya often said that a smooth release is "boring," meaning well-planned and low drama. By following a checklist, doing rehearsals (smoke tests), and communicating every step, they avoided crisis-mode panic. Team Vertex now lives by the mantra that a hurried or impulsive launch leads to chaos, whereas a well-orchestrated launch feels routine.

Deployment is a team sport: The MVP went live because everyone played their part. This wasn't a one-person show. QA, developers, PO, SM, UX, and support all worked together. They held each other accountable: Malik ran tests, Alex handled servers, Maria informed stakeholders, Jenna checked the UI, and Maya kept the train on its tracks. When things worked, they knew it was thanks to their cross-functional teamwork. They learned that finger-pointing only kills momentum; shared ownership, on the other hand, builds success.

Documentation drives clarity: Keeping notes and writing things down turned out to be invaluable. The release notes Maria sent were referenced by everyone (marketing posted them on Slack!). The changelog helped the team remember what changed between versions. Post-release, their written postmortem captured ideas while they were fresh. Good documentation prevented confusion as everyone had the same information about what was deployed and why. The team vows that next time, they'll document even a little more (for example, their emergency rollback steps) because clarity reduces fire drills.

Value delivery starts at release: Throughout the process, the team repeatedly reminded themselves that no real value is delivered until users actually have the software. They had fun building features, but they only added meaning when people could use the app. As Maya put it: *"We've been building the plane, but we haven't landed it until now."* This reinforced to Team Vertex that planning and coding are steps on the path, but the destination is the release itself. Their journey to Agile success is measured in those moments when users can finally pick up the app in production.

Team Vertex walks away from this launch more confident. They saw that following Scrum values (collaboration, clarity, and iterative delivery) applies not just to development work but to releases too. These lessons, some practical, some philosophical, will guide them as they prepare for their next Sprint and beyond. The PetTracker MVP is now live, and the real adventure of learning from real users has just begun.

PURPOSE OF THE RETROSPECTIVE

Two weeks after production launch, Team Vertex came together for a formal **MVP Retrospective**. This session wasn't just about celebrating wins: it was about converting experience into repeatable success patterns and refining what didn't work.

Maya opened the retro by stating:

"We've shipped a working product. Now let's shape how we grow from here."

KEY MVP METRICS

Metric	Value
Sprint Velocity (Average)	17 story points
Total Stories Delivered	24
UAT Bugs Reported Post-Deploy	3 minor, 0 critical
App Store Crash Rate	0.0% (within first 14 days)

Feature Usage (Top 3)	Walk Log > Feeding Log > Profile Edit
Time-to-Hotfix (avg)	< 48 hours

EXPANDED CONTEXT ON EACH KEY METRIC

1. Sprint Velocity (Average: 17 story points)

Velocity demonstrated the team's sustainable pace across Sprints. By anchoring expectations around ~17 points, the team now had a baseline for forecasting future capacity and release planning.

2. Total Stories Delivered (24)

This showed the scope delivered versus what was planned. Twenty-four completed stories confirmed that the MVP delivered meaningful, user-facing functionality rather than half-finished work.

3. UAT Bugs Reported Post-Deploy (3 minor, 0 critical)

Low bug counts validated the effectiveness of QA and developer collaboration. The absence of critical defects meant the MVP maintained trust with users from day one.

4. App Store Crash Rate (0.0% in first 14 days)

A zero-crash launch is rare and signaled engineering excellence. This helped strengthen early adopter confidence and provided a marketing talking point.

5. Feature Usage (Top 3: Walk Log > Feeding Log > Profile Edit)

Usage data was vital in shaping the backlog. Walk Log's dominance revealed where the app was providing the most value, helping prioritize enhancements in future Sprints.

6. Time-to-Hotfix (avg < 48 hours)

Responsiveness after launch was equally important as the features delivered. The ability to patch issues within two days demonstrated agility and commitment to user experience.

How the Metrics Fueled Growth Discussions

Instead of treating the numbers as a scoreboard, Maya guided the team to ask:

- *What patterns here should we preserve?* (sustainable velocity, fast hotfix turnaround)

- *What signals guide our next priorities?* (feature usage insights)
- *Where do we need to tighten feedback loops?* (continue minimizing UAT bugs)

This allowed the retrospective to shift from a **"What happened" conversation** into a **"How do we grow smarter?" playbook.**

PRE-PRODUCTION DEPLOYMENT CHECKLIST

Environment & Code Readiness

- ☐ Verify all code changes have been merged into the release branch.
- ☐ Confirm successful build pipeline (CI/CD) with no errors.
- ☐ Ensure automated unit, integration, and regression tests pass.
- ☐ Validate feature toggles or flags are configured correctly.
- ☐ Confirm version number/tag matches release documentation.

Infrastructure & Configuration

- ☐ Check pre-prod environment mirrors production (DB schema, APIs, integrations).
- ☐ Verify secrets, environment variables, and configs are up to date.
- ☐ Ensure rollback strategy and scripts are documented and tested.
- ☐ Validate monitoring and logging tools are active in pre-prod.

Approvals & Communication

- ☐ Obtain stakeholder sign-off (Product Owner, QA Lead, Release Manager).
- ☐ Notify support and operations teams of deployment window.
- ☐ Finalize release notes and update knowledge base.

Deployment Checklist (During Release)

Execution

- [ ] Trigger deployment pipeline (automated or manual).
- [ ] Monitor build and deployment logs in real time.
- [ ] Validate DB migrations or schema changes apply successfully.
- [ ] Ensure feature flags are enabled/disabled as intended.

Smoke Testing

- [ ] Confirm application starts without errors.
- [ ] Run critical path smoke tests (login, transactions, APIs).
- [ ] Validate health checks return "OK" across all services.

Communication

- [ ] Provide real-time status updates to stakeholders.
- [ ] Log deployment start and end times.

POST-DEPLOYMENT CHECKLIST

Validation

- [ ] Confirm user journeys work as expected in production.
- [ ] Run post-deployment regression suite.
- [ ] Verify monitoring dashboards show normal traffic/response times.
- [ ] Check error logs for unexpected issues.

Data Integrity

- [ ] Validate database updates with sample queries.
- [ ] Ensure no data loss or corruption occurred.

Documentation & Communication

- ☐ Publish release notes to end users/stakeholders.
- ☐ Document any hotfixes, manual steps, or deviations.
- ☐ Update runbooks with lessons learned.

Retrospective & Rollback Readiness

- ☐ Conduct a quick deployment retrospective (what went well, what to improve).
- ☐ Keep rollback plan ready for 24–48 hours post-deployment.

Exercises & Activities

Production Deployment Exercise

Title: MVP Release Readiness — The Final Confidence Check

Export: 08_ProdDeployment_Evidence.pdf

Objective:

Show that your build is ready for real users by proving the team can plan, execute, and validate a safe production launch.

Your Mission

Complete the four-step deployment readiness package:

Release Confidence Checklist

Fill out a short checklist confirming:

Checklist Item	Status	Checklist Item
Code merged to release branch	☑ / ✕	
Build pipeline succeeded	☑ / ✕	Include screenshot
Feature toggles correct	☑ / ✕	Which features are ON?

UAT Smoke tested	✅ / ✕	Tester initials
Rollback plan confirmed	✅ / ✕	1-sentence rollback action
Stakeholders notified	✅ / ✕	Provide screenshot of message
Monitoring alert active	✅ / ✕	Screenshot or description

2. Role-Based Deployment Plan

Define who does what on launch day:

Role	Role	Role
Product Owner	You	Release notes + go/no-go
Scrum Master		Facilitate launch
Developer		Merge + trigger pipeline
QA		Post-deploy smoke test
Support Rep		Ready for user feedback

📌 5 roles required

📌 One sentence per responsibility

3. Stakeholder Release Announcement

Write the actual communication. Keep it short, professional, and confident:

Subject: 🚀 PetTracker MVP is Live!

Version: 1.0

Released: <today's date>

Now Available: ✅ Sign Up, ✅ Login, ✅ Pet Profile

Early Release Notes: <1–2 bullets>

Support Contact: <your contact>

Paste screenshot of where you would publish this (Teams, Email, Confluence)

4. Prod Smoke Test Log

Prove the app works after deployment:

Test Case	Status	Evidence
Sign up with valid email	☑	Screenshot
Login after sign up	☑	Screenshot
Create pet profile	☑	Screenshot
Profile persists after relogin	☑	Screenshot

📌 Include 4 small screenshots and timestamps = authenticity

Deliverable Format

Upload as a 4–6 page PDF:

Page Structure

1. Title Page — PetTracker MVP Release
2. Release Checklist + Screenshots
3. Role-Based Deployment Plan
4. Stakeholder Announcement Proof
5. Prod Smoke Test Evidence
6. (Optional) Learnings / Risks / Next iteration

Filename: 08_ProdDeployment_Evidence.pdf

Portfolio Badge Earned

"Production-Ready Release Planner"

This exercise tells hiring managers:

☑ You can ship.

☑ You can communicate.

☑ You can protect product quality in the real world.

20. BURN CHARTS & REAL-TIME ADJUSTMENT

WHY BURN CHARTS MATTER BEYOND TRACKING

Burn charts aren't just lines on a graph: they serve different purposes for different roles, making them a shared conversation tool rather than a Scrum Master's reporting mechanism.

- For the Team: Burn charts help members self-manage their pacing. They can see if work is bunching up at the end of the Sprint, if tasks are stalling, or if progress is flowing smoothly. This allows them to adjust without waiting for someone else to point it out.
- For the Product Owner: Burn charts reveal when scope shifts or unexpected challenges may require stakeholder communication. A flat line or sudden spike can trigger a conversation about trade-offs, ensuring transparency and trust with the business side.
- For Leadership: Burn charts illustrate predictability over time. Consistent patterns signal a maturing team; volatile patterns highlight where support, coaching, or process improvements may be needed. Leaders can use these insights to guide long-term planning without micromanaging the team.

By framing burn charts this way, Team Vertex transformed them from a progress-tracking tool into a teamwide feedback loop, helping everyone align on delivery pace, expectations, and growth.

TEAM VERTEX IN ACTION

In Sprint 3, midway through building the Walk Log feature, the burn down chart showed progress flattening unexpectedly. Instead of waiting until the retrospective, Maya called it out during Daily Scrum:

"This chart is telling us something; let's dig in. Are we blocked, or did scope just grow?"

The team quickly uncovered that GPS permission handling was more complex than expected, and tasks were piling up in *In Progress*. Jamal suggested spiking the problem, while Maria updated the Product Owner so stakeholders wouldn't be surprised by a late adjustment.

By using the burn chart as a **conversation starter rather than a scoreboard**, Team Vertex not only unblocked their work but also reinforced trust with leadership, who could see the team adapting in real time.

BURN CHART ANTI-PATTERNS: WHAT NOT TO DO

Burn charts are powerful when used for insight, but misused, they can distort behaviors and harm team morale. Here are common anti-patterns to watch for:

- **The Waterfall Sprint (Flat Line Until the End):** If the burn down line stays flat for most of the Sprint and drops sharply in the last two days, it signals that work is being held back until the end, mimicking waterfall delivery. This hides risks until it's too late and often leads to rushed, low-quality work.
- **The Scope Creep Spiral (Burn Up Keeps Growing):** A burn up chart with the total scope climbing every few days shows that new requirements are constantly being added mid-Sprint. The result? The finish line keeps moving, and morale drops as the team feels they're "never done."
- **The Blame Game (Charts as Weapons):** Burn charts should spark conversations, not finger-pointing. If a chart is used to single out underperformers or "prove" that a team is failing, psychological safety erodes. The team stops being honest about blockers and instead focuses on avoiding blame.

Lesson from Team Vertex

During Sprint 2, Team Vertex noticed their burn down chart was jagged, with long flat zones. Instead of blaming QA for "slowing things down," Maya reframed it:

"The chart's not about who's behind: it's showing us where our process needs strengthening. Let's use this as a learning moment."

That discussion led to tighter Acceptance Criteria and a reusable UI validation checklist. The anti-pattern became an opportunity to improve.

Key Reminder: Burn charts are mirrors, not weapons. They reflect patterns that teams can learn from, but it's the conversation, not the curve, that drives improvement.

BURN CHARTS AND THE SCRUM PILLARS

Burn charts aren't just project visuals: they embody the three pillars of Scrum: **transparency, inspection, and adaptation.**

- **Transparency:** Burn charts make progress (or lack of it) visible to everyone. Instead of hidden delays or silent blockers, the whole team, *and* stakeholders, can see the reality of where the Sprint stands.
- **Inspection:** By reviewing the chart daily, the team inspects how their actual progress compares to the plan. Flat lines, sharp drops, or scope changes all become signals worth examining.
- **Adaptation:** The power of a burn chart lies in how the team responds. If scope grows unexpectedly or progress stalls, the team adapts: by cutting scope, spiking a blocker, or shifting priorities to protect the Sprint goal.

Team Vertex in Practice

In Sprint 1, when the burn down chart stayed flat through most of the Sprint, the team inspected their process and realized work was piling up in *In Progress*. Instead of pushing harder in silence, they adapted: time-boxing story discussions and breaking large tasks into smaller slices.

By Sprint 4, their burn charts reflected a smoother, steady decline, proof that the Scrum pillars weren't abstract principles, but daily habits fueling team maturity.

BEYOND BURN CHARTS: BUILDING A SPRINT HEALTH DASHBOARD

While burn down and burn up charts offer a clear view of progress, mature Agile teams often expand their lens. By combining different metrics, they create a **Sprint Health Dashboard**: a holistic picture of speed, quality, flow, and team well-being.

- **Cumulative Flow Diagram (CFD):**

  The CFD shows how work moves across the board: backlog, in progress, testing, and done. If one band balloons (like "In Testing"), it signals a bottleneck. Team Vertex used this in Sprint 3, when too many stories piled up waiting for QA review. Seeing the jam visually helped them swarm on testing and rebalance their effort.

- **Escaped Defects (Post-Sprint Bugs):**

  Tracking bugs that slip into production keeps quality visible. In Sprint 2, Vertex saw rework caused by missing form validations. By counting escaped defects, they were able to justify investing in reusable checklists, reducing post-deploy surprises.

- **Lead Time & Cycle Time:**

  Lead time measures from story creation to delivery.

  Cycle time measures from "in progress" to "done."
  These metrics reinforce efficiency beyond just story points. In Sprint 4, Maria noticed that although velocity stayed steady, cycle time dropped: proving the team was finishing work faster, not just doing the same amount.

- **Work Item Age:**

  This tracks how long tasks linger in progress. If a story sits open for days, it may be blocked or too large. Vertex used this metric to spot hidden blockers in their GPS permissions story (Sprint 3), prompting them to break it into smaller spikes and features.

- **Team Happiness / Morale Metric:**

  Charts measure delivery, but the human side matters too. Some teams run a simple retro poll ("How satisfied were you this Sprint? 1–5"). In Sprint 2, when bug rework caused stress, Maya added a quick pulse check. Scores dipped, sparking a conversation that led to pairing developers with QA earlier in the Sprint.

Takeaway: Burn charts tell one part of the story, but pairing them with additional metrics gives a full dashboard of delivery health. Together, they help teams balance *speed, quality, flow,* and *well-being*: the real ingredients of sustainable agility.

Sprint Health Dashboard: Key Metrics

Metric	What It Shows	Team Vertex Example
Burn Down / Burn Up Charts	Sprint pacing and scope alignment.	Sprint 3 burn down exposed a GPS permissions blocker, leading to a scope adjustment.
Metric	**What It Shows**	**Team Vertex Example**
Cumulative Flow Diagram (CFD)	Flow of work across stages; highlights bottlenecks.	Sprint 3 showed too many items stuck in Testing, prompting swarming on QA.
Escaped Defects	Quality measure: bugs missed during Sprint testing.	Sprint 2 form validation bugs revealed weak Acceptance Criteria; checklists fixed this.
Lead Time & Cycle Time	Efficiency of delivering stories from creation → done.	Sprint 4 cycle time dropped, proving smoother workflow despite steady velocity.
Work Item Age	How long items stay open; surfaces hidden blockers.	GPS story sat in In Progress too long: team broke it into smaller spikes.
Team Happiness / Morale	Team sentiment; signals stress or satisfaction.	Sprint 2 morale dip during bug rework triggered earlier Dev–QA pairing.

BURN DOWN CHART SNAPSHOTS (SPRINT 1–4)

Sprint 1: Profile & Login

- **Trend:** Slight dip early, major progress in final 2 days.
- **Insight:** Work was back-loaded. The team overestimated how long Firebase Auth integration would take.
- **Adjustment:** Added time-boxing to user story discussions during Sprint Planning.

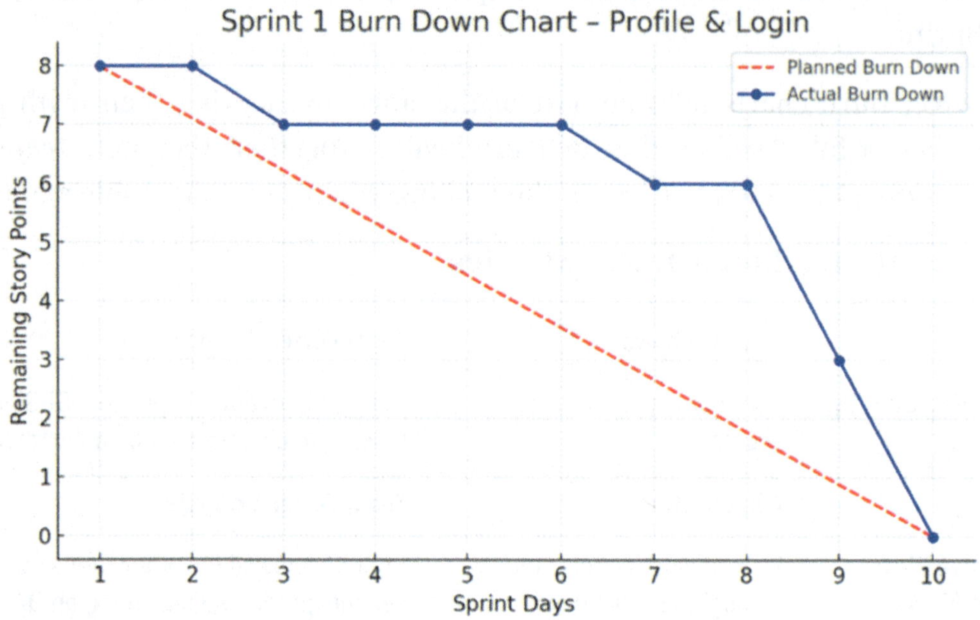

Sprint 2: Feeding Log

- **Trend:** Jagged decline with multiple flat zones.
- **Insight:** QA surfaced bugs that caused task rework mid-Sprint.
- **Adjustment:** Added stronger Acceptance Criteria and built a reusable UI form validation checklist.

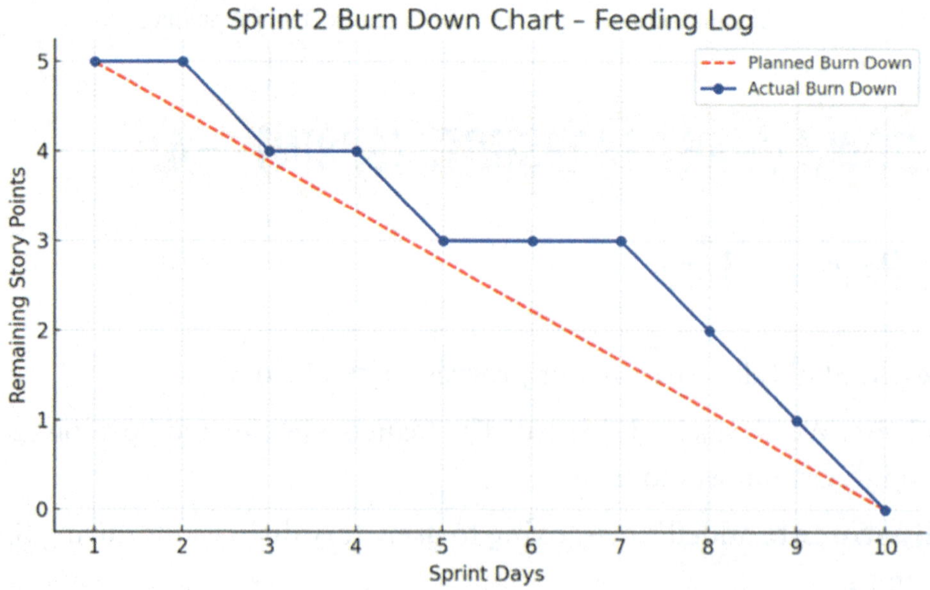

Sprint 3: Walk Log with GPS

- **Trend:** Smooth progress until Day 5, then sharp scope cut.
- **Insight:** GPS permission handling proved more complex than expected.
- **Adjustment:** Spiked feature feasibility during planning and added fallback logic for denied permissions.

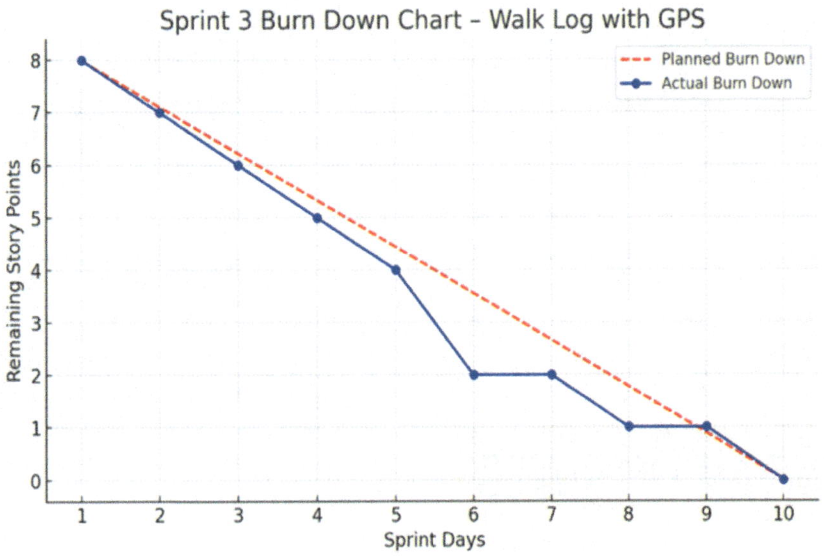

Sprint 4: Medication Logging

- **Trend:** Consistent burn down with minimal rollbacks.
- **Insight:** By this Sprint, the team's velocity and coordination had matured.
- **Adjustment:** Used MedicationLog as a form component template for future logs.

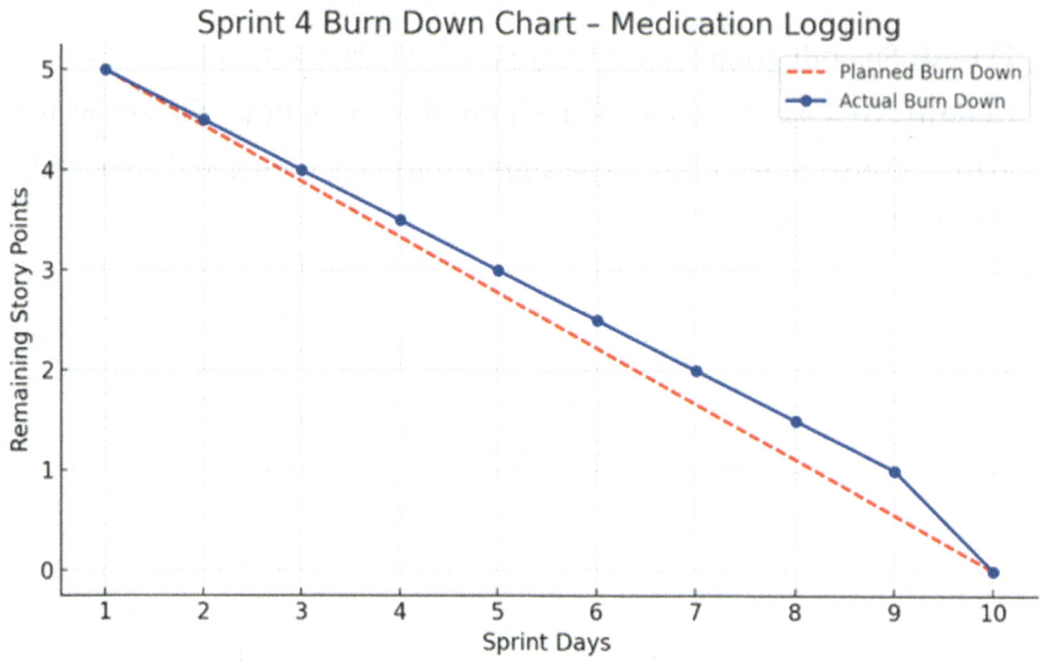

HOW THE TEAM IMPROVED

Area	Sprint 1	Sprint 4 (Improved)
Acceptance Criteria	Vague, missing edge cases	Clearly defined with validation details
Testing	Manual only	Added reusable test cases and field checks
Estimation	Based on guesswork	Based on past velocity and spike results
Design handoff	Minimal walkthroughs	Detailed sessions with tagged mockups
QA bugs	High volume	Reduced through reusable patterns

Velocity Snapshot

Sprint	Story Points Committed	Story Points Completed	Completion Rate
1	8	5	63%
2	5	5	100%
3	8	6	75%
4	5	5	100%

Team Reflection

In their Sprint 4 Retrospective, Maria noted:

"Our charts now reflect the maturity of our process, not just task completion. We course-correct sooner, test faster, and align earlier."

KEY TAKEAWAYS

- Charts aren't just data; they tell a story.
- Use burn patterns to spark conversations in Retrospectives.
- Refine the Definition of Ready and Definition of Done based on what patterns repeat.
- Visualize progress as a coaching tool, not just for reporting.

Exercises & Activities

Exercise: Read the Chart, Shape the Sprint

Export name: 15_BurnCharts_InsightActivity.pdf. This exercise trains you to interpret burn charts as early-warning signals and drive mid-Sprint adaptation like a seasoned Scrum Master.

Part 1 — Interpret the Sprint Pattern

Below are four burn down chart scenarios. For each chart, answer:

A) What story is this chart telling?

B) What conversation would you initiate with the team?

C) What action would you help the team take tomorrow?

1. The Cliff Dive

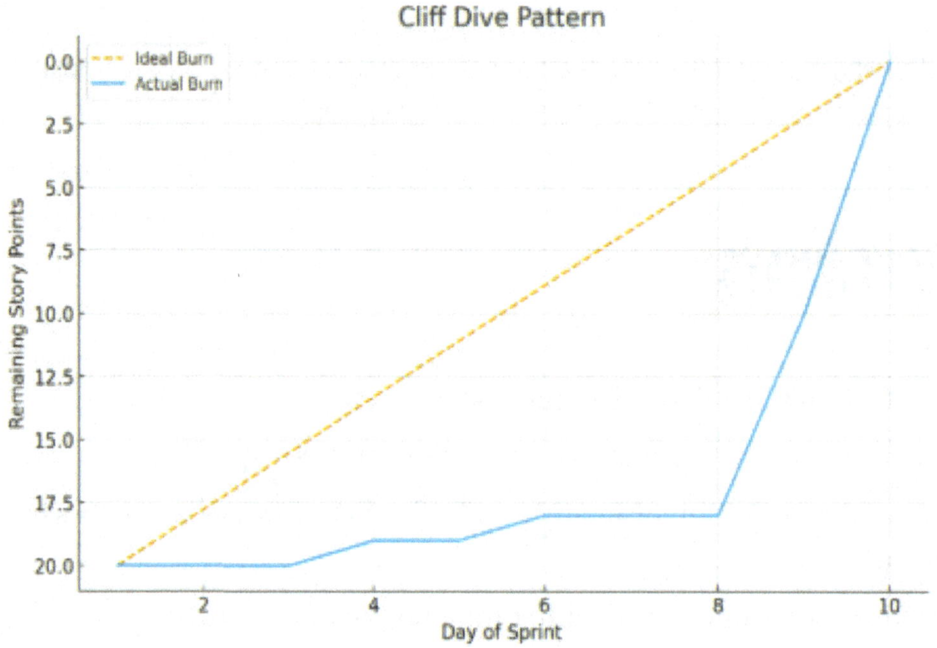

Progress flat until the last 2 days → sudden drop → What does this say about workflow + transparency?

2. The Scope Creeper

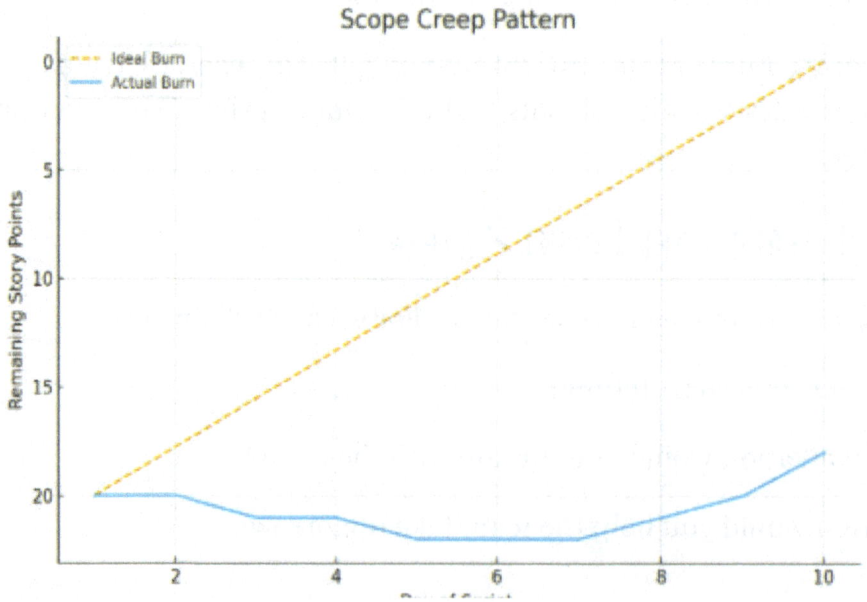

Burn up line keeps rising → scope gets added mid-Sprint → How do you protect Sprint Goal + morale?

3. The Jagged Saw

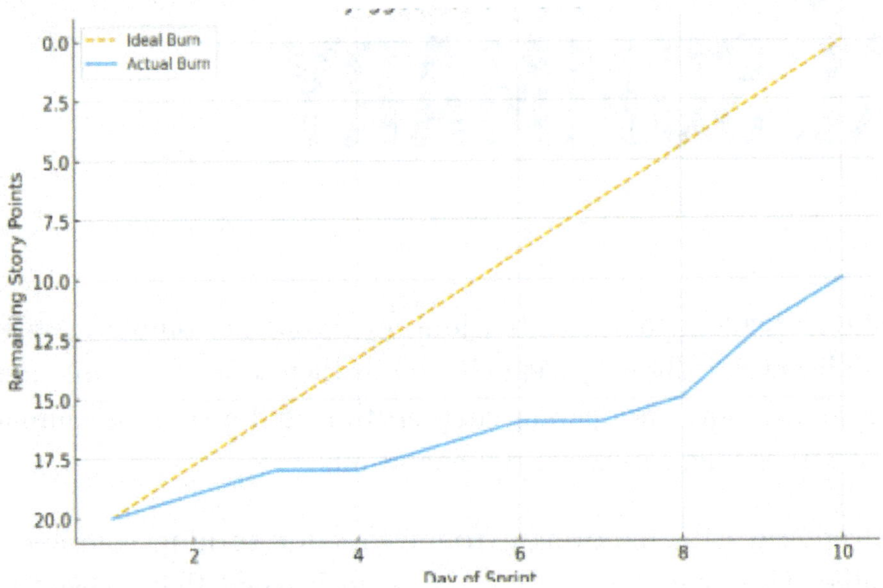

Up-and-down rework patterns → What does this signal about quality and acceptance criteria?

4. The Smooth Decline

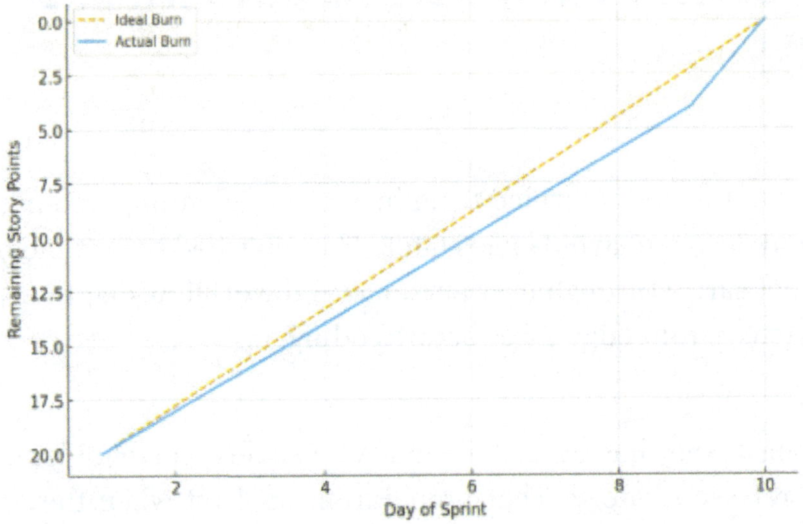

Consistent pace + clean testing → What habits enabled this maturity? How do you lock them in?

Write: 2–3 bullet points per section.

21. FINAL SUMMARY

Scrum Quest took readers on a narrative journey through a complete Agile software-development life cycle, following Team Vertex as they conceived, built, and released their PetTracker application. The opening chapters framed the SDLC as a human-centric, iterative process rather than a series of rigid hand-offs.

Scrum was introduced as a lightweight framework for turning complex work into incremental value. Throughout the story, the team learned that success wasn't just about process; it was about people. They started with a clear product vision, crafted a realistic scope and risk-mitigation plan, and formed a cross-functional unit committed to working together.

LAYING THE FOUNDATION: VISION, ALIGNMENT, AND ARCHITECTURE

- During Sprint 0, the newly formed team invested in planning, requirements elicitation, and design to avoid chaos later. They clarified the problem space ("forgetting pet care") through user personas and workflow diagrams, which forced them to confront edge cases before coding.

By the end of this phase, they had agreed on an MVP backlog, stakeholder engagement plan, and iterative release strategy. That preparation paid off when they moved into development: the team set up the technical architecture, chose their tool stack, and established definitions of "Ready" and "Done." This foundation underscored a key Scrum lesson: transparency and alignment reduce re-work and build trust.

ITERATIVE DELIVERY AND LEARNING

Subsequent Sprints demonstrated how small, vertical slices of functionality could deliver value quickly while exposing real-world challenges.

- Sprint 1 introduced user authentication and pet profiles, and a duplicate-email bug taught the value of clear acceptance criteria and QA involvement early in planning.
- In Sprint 2, the "simple" feeding log revealed hidden complexity around validation and user feedback, forcing the team to refine their definition of Done.
- The Walk Log in Sprint 3 brought technical hurdles, like GPS permissions and background timers, that were overcome through pair programming, rapid prototyping, and continuous testing.
- Sprint 4 added medication logging and scheduling, where a crash caused by a missing time field led to better default handling and more robust input validation.

Across all Sprints, the team held daily Scrums, Sprint reviews, and retrospectives to inspect progress and adapt plans. They learned that burn-down and burn-up charts weren't just reports; they triggered conversations about scope, velocity, and quality.

The team's velocity stabilized as they refined their estimations based on empirical data, reflecting Scrum's emphasis on empiricism. Retrospectives uncovered patterns, such as over-commitment leading to burnout, that were addressed by renegotiating scope and protecting sustainable pace, echoing Scrum's focus on inspection, adaptation, and respect.

BEYOND MECHANICS: EMBRACING VALUES AND LEADERSHIP

While Team Vertex mastered ceremonies and artifacts, their real growth came from adopting Scrum values. They learned that great teams did not avoid conflict; they had the courage to raise issues and use retrospectives as mirrors. Backlogs became outcome-driven rather than task lists; user stories were prioritized for customer value, not developer convenience. Leaders stopped micromanaging and instead cultivated flow, encouraging the team to self-organize. The Scrum Guide reminds us that commitment, focus, openness, respect, and courage guide behavior: values that Team Vertex gradually embodied.

Later chapters moved beyond the MVP to address leadership challenges: conflict resolution, stakeholder management, and risk mitigation. Techniques like neutral framing, the "5 Whys," and behavioral contracting equipped the Scrum Master to coach through tension and keep the team aligned.

A final review Sprint showed how experienced teams integrated new technical initiatives (IoT sensors, performance tuning) while maintaining sustainable velocity and adapting to shifting stakeholder priorities. This section reinforced that mature Scrum is less about rituals and more about fostering a resilient, self-managing team.

THE BIG PICTURE

By the end of Scrum Quest, Team Vertex had delivered a working product and transformed their way of working. They began as a group of individuals and finished as a cohesive, empowered team capable of navigating complexity.

Readers came away with concrete techniques, personas, backlog refinement, acceptance-criteria checklists, burn charts, and, more importantly, with a mindset that values continuous learning and human-centric leadership. In line with Scrum theory, the story proved that empiricism, transparency, inspection, and adaptation are not abstract concepts but daily practices that help teams deliver real value. Great teams, it concluded, "know that every backlog tells a story, every Sprint is a test, every retro is a mirror, and every team is a work in progress."

22. FROM MASTERY TO MOMENTUM

CELEBRATE THE JOURNEY

You've made it. From the first spark of an idea in the Agile Project Scope to a working MVP in production, you've walked every step with Team Vertex. You witnessed their Sprint struggles, their user story negotiations, their retrospectives, their metrics mishaps, and their growth into a truly cross-functional Agile team. But more than that, you journeyed through your own learning arc. You learned and lived Scrum.

You now know that every backlog tells a story. Every Sprint is a test. Every retro is a mirror. And every team is a work in progress.

Whether you're a new Scrum Master or a seasoned Product Owner, this book has shown that:

- Great teams don't avoid conflict. They **inspect and adapt** through it.
- Great backlogs don't just hold tasks. They **prioritize outcomes**.
- Great leaders don't micromanage. They **cultivate flow**.

And the best teams understand that done isn't just when the code is shipped. It's when users see the value, teams find their groove, feedback loops are tight, and trust runs deep.

PAUSE AND REFLECT

Before you leap into the next phase of your career or agile journey, ask yourself:

- Which part of Team Vertex's story mirrored your own experience?
- What new rituals or metrics are you ready to try with your team?
- How will you measure success beyond story points?
- What can you bring back to your organization as knowledge and **practice**?

WHAT'S NEXT?

Team Vertex's journey shows that mastery isn't a finish line: it's the launchpad for sustained momentum. The practices they refined, from burn charts to retrospectives, proved that agility is built through small, repeatable habits. To help you apply these lessons in your own context, the next chapter equips you with a **practical toolkit**, a collection of templates, checklists, and quick-reference guides designed to turn ideas into daily action.

23. KEY DEFINITIONS & GLOSSARY

AGILE & MINDSET FOUNDATIONS

AGILE

Agile is an approach to product and software delivery that prioritizes learning through small releases, frequent feedback, and continuous improvement. Instead of trying to predict everything upfront, Agile teams adapt as they learn and focus on delivering real value early and often.

AGILE MINDSET

The Agile mindset is a way of thinking that values flexibility, collaboration, transparency, and learning over rigid plans and control. It is less about following steps and more about how teams respond when reality changes, which it always does.

AGILE MANIFESTO

The Agile Manifesto is the foundational statement that defines what Agile values most. It outlines four core values and twelve principles that guide how Agile teams work together and deliver value.

AGILE VALUES

Agile values describe what matters most when making decisions:

- Individuals and interactions

- Working software
- Customer collaboration
- Responding to change

These values help teams prioritize people, outcomes, and adaptability over heavy process and rigid plans.

AGILE PRINCIPLES

The Agile Principles are twelve guiding statements that explain how Agile values show up in real work. They emphasize early delivery, collaboration, sustainable pace, technical excellence, and regular reflection to improve.

EMPIRICISM

Empiricism is the idea that decisions should be based on what is known through observation and experience, not assumptions. In Agile and Scrum, teams rely on transparency, inspection, and adaptation to guide their next steps.

INSPECT AND ADAPT

Inspect and Adapt is the practice of regularly reviewing outcomes and adjusting behavior based on what is learned. Instead of sticking to a plan that no longer makes sense, teams pause, evaluate, and improve continuously.

ITERATIVE DEVELOPMENT

Iterative development means building a product in small cycles, learning from each iteration, and improving the next one. Each iteration adds insight, not just features.

INCREMENTAL DELIVERY

Incremental delivery is the practice of delivering usable pieces of the product over time rather than waiting for one large release. Each increment provides value and reduces risk.

VALUE DRIVEN DELIVERY

Value driven delivery focuses the team on solving real user problems and delivering outcomes that matter, not just completing tasks or checking boxes. Work is prioritized based on impact rather than effort alone.

CROSS FUNCTIONAL TEAM

A cross functional team includes all the skills needed to deliver a product increment, such as design, development, testing, and analysis, without relying on handoffs to external groups. The team succeeds or fails together.

SUSTAINABLE PACE

Sustainable pace means working at a speed the team can maintain over time without burnout. Agile teams prioritize long term effectiveness over short term heroics.

SCRUM FRAMEWORK CORE CONCEPTS — DEFINITIONS

SCRUM FRAMEWORK

The Scrum Framework is a lightweight structure that helps teams deliver complex products through short, focused cycles called Sprints. It provides clear roles, events, and artifacts that create transparency, accountability, and a steady rhythm for learning and delivery.

THE SCRUM ROLES

SCRUM TEAM

The Scrum Team is a small, self managing group responsible for delivering value every Sprint. It includes the Product Owner, Scrum Master, and Development Team, all working together toward a shared goal.

PRODUCT OWNER

The Product Owner is responsible for maximizing product value. They define what the team should build and why, manage and prioritize the Product Backlog, and represent the needs of customers and stakeholders.

SCRUM MASTER

The Scrum Master is a servant leader who helps the team use Scrum effectively. They coach the team, remove blockers, protect the Sprint focus, and help the organization support Agile ways of working.

DEVELOPMENT TEAM

The Development Team is a cross functional group that designs, builds, tests, and delivers the product Increment. They decide how the work gets done and are collectively accountable for quality and delivery.

THE 5 SCRUM EVENTS

SPRINT

A Sprint is a fixed length time period, typically one to four weeks, during which the team delivers a usable product Increment. Each Sprint includes planning, execution, review, and reflection.

SPRINT GOAL

The Sprint Goal is a short statement that explains the purpose of the Sprint. It provides focus and helps the team make decisions when tradeoffs are needed during the Sprint.

SPRINT PLANNING

Sprint Planning is the event where the Scrum Team decides what work will be done in the upcoming Sprint and how it will be achieved. The team selects items from the Product Backlog and commits to a shared Sprint Goal.

DAILY SCRUM

The Daily Scrum is a fifteen minute team check in held every day of the Sprint. The team inspects progress toward the Sprint Goal and adjusts their plan as needed.

SPRINT REVIEW

The Sprint Review is held at the end of the Sprint to demonstrate completed work to stakeholders. Feedback is gathered and the Product Backlog is updated based on what was learned.

SPRINT RETROSPECTIVE

The Sprint Retrospective is a dedicated time for the team to reflect on how they worked together. The team identifies what went well, what did not, and what improvements to try next.

THE 3 SCRUM ARTIFACTS

PRODUCT BACKLOG

The Product Backlog is an ordered list of everything that could be built for the product. It is the single source of truth for future work and is owned and prioritized by the Product Owner.

SPRINT BACKLOG

The Sprint Backlog contains the work selected for the current Sprint along with the team's plan to deliver it. It represents the team's commitment to achieving the Sprint Goal.

INCREMENT

The Increment is the usable, working product produced during a Sprint. It must meet the Definition of Done and be ready for release, whether or not it is actually released.

SDLC CONCEPTS (AGILE-ALIGNED)

SOFTWARE DEVELOPMENT LIFE CYCLE

The Software Development Life Cycle, or SDLC, is the structured set of phases teams move through to plan, build, test, release, and maintain software. In Scrum Quest, the SDLC is treated as a continuous cycle that repeats within each Sprint rather than a one time linear process.

AGILE ALIGNED SDLC

The Agile aligned SDLC is the application of traditional SDLC phases within an Agile framework. Instead of completing each phase once, teams revisit planning, design, development, testing, and release in small cycles, allowing for faster feedback and adaptation.

MINI SDLC

A Mini SDLC refers to completing all SDLC phases within a single Sprint. Each Sprint becomes a full delivery cycle where planning, design, development, testing, and review happen together to produce a usable product Increment.

SOFTWARE DEVELOPMENT LIFE CYCLE PHASES

INITIATION AND PLANNING

Initiation and Planning is the phase where the product vision, goals, scope, risks, and high level approach are defined. In Scrum, this work often begins in Sprint Zero and continues as the backlog evolves.

REQUIREMENTS ANALYSIS

Requirements Analysis is the phase where product needs are clarified and broken down into actionable work. In Scrum, this happens through backlog refinement, user story writing, and defining acceptance criteria.

DESIGN

Design is the phase where solutions are visualized and planned before development begins. In Agile delivery, design is done just in time and collaboratively, often within the same Sprint as development.

DEVELOPMENT

Development is the phase where the team builds the product by writing code and integrating components. Work is completed incrementally and aligned to the Sprint Goal.

TESTING

Testing is the phase where the team verifies that the product works as intended and meets quality standards. In Scrum, testing happens continuously during the Sprint, not just at the end.

DEPLOYMENT AND RELEASE

Deployment and Release is the phase where completed work is prepared for delivery to users. This may include deploying to test or production environments and ensuring the product is ready for real world use.

MAINTENANCE

Maintenance is the phase where the team monitors the product, fixes issues, and improves performance after release. Feedback from users and systems informs future backlog items.

SPRINT READINESS & QUALITY GUARDRAILS

SPRINT ZERO

Sprint Zero is a preparatory Sprint used to establish clarity, alignment, and technical readiness before feature development begins. It focuses on vision, backlog setup, design alignment, environment readiness, and quality standards rather than delivering end user features.

DEFINITION OF READY

The Definition of Ready is a shared checklist that determines whether a backlog item is clear and complete enough to be pulled into a Sprint. It typically includes a clear description, business value, acceptance criteria, and resolved dependencies.

DEFINITION OF DONE

The Definition of Done is a set of agreed upon criteria that defines when work is truly complete. It ensures that each Increment meets quality, testing, and usability standards before it is considered finished.

ACCEPTANCE CRITERIA

Acceptance Criteria define the conditions that must be met for a user story to be accepted. They clarify expected behavior, remove ambiguity, and provide a shared understanding of what success looks like.

GHERKIN FORMAT

Gherkin format is a structured way of writing acceptance criteria using simple, scenario based language. It describes system behavior through clear conditions and outcomes that are easy for both technical and non technical team members to understand.

GIVEN WHEN THEN

Given When Then is the sentence structure used in Gherkin format to define acceptance scenarios. It describes the starting context, the action taken, and the expected result in a clear and testable way.

INVEST

INVEST is a guideline used to evaluate whether a user story is well written and ready for development. It ensures stories are independent, negotiable, valuable, estimable, small, and testable.

INVEST works best when paired with the 3 C Model. The Card captures the core intent of the story, Conversation allows the team to explore details, risks, and edge cases together, and Confirmation provides acceptance criteria that define when the story is complete.

Together, INVEST and the 3 C Model help teams avoid treating user stories as fixed specifications. Instead, stories remain flexible, collaborative, and focused on delivering clear, testable value.

SPRINT READINESS REVIEW

The Sprint Readiness Review is a checkpoint used to confirm that upcoming backlog items meet the Definition of Ready before Sprint Planning. It helps teams reduce risk, avoid last minute surprises, and commit with confidence.

PRODUCT & BACKLOG MANAGEMENT

PRODUCT VISION

The Product Vision is a short, clear statement that explains why the product exists and who it is for. It guides decision making, prioritization, and tradeoffs throughout the product lifecycle.

PRODUCT GOAL

The Product Goal describes a measurable outcome the product aims to achieve over time. It provides focus for backlog prioritization and helps the team understand what success looks like beyond individual Sprints.

ROADMAP

A roadmap is a high level view of where the product is headed and the sequence of outcomes the team plans to deliver. In Agile, roadmaps are flexible and evolve as teams learn from feedback.

AGILE PROJECT SCOPE

Agile Project Scope defines what is included and excluded from the current product initiative. It sets boundaries without locking the team into rigid plans, helping manage expectations and reduce scope creep.

MVP

MVP, or Minimum Viable Product, is the smallest version of a product that delivers real value to users and allows the team to learn from feedback. It focuses on outcomes rather than completeness.

MMP

MMP, or Minimum Marketable Product, is a more complete version of the product that is ready for broader release or commercialization. It builds on the MVP by adding features needed for market readiness.

EPIC

An Epic is a large body of work that represents a significant product capability or user outcome. Epics are broken down into smaller features and user stories over time.

FEATURE

A Feature is a distinct piece of functionality that delivers value to users. Features often span multiple user stories and help organize the backlog around outcomes rather than tasks.

USER STORY

A User Story is a short description of functionality written from the perspective of the user. It captures who the user is, what they need, and why it matters, while leaving room for collaboration and discovery.

TASK

A Task is a specific piece of work required to complete a user story. Tasks are owned by the team and represent how the work will be done.

STORY SLICING

Story slicing is the practice of breaking large stories into smaller, manageable pieces that can be completed within a single Sprint. Good slicing focuses on delivering value incrementally rather than dividing work by role.

PRIORITIZATION

Prioritization is the process of ordering backlog items based on value, risk, and learning impact. The goal is to ensure the most important work is done first.

ESTIMATION

Estimation is the practice of assessing the relative size or effort of backlog items. In Agile, estimation is used for planning and forecasting, not for measuring individual performance.

STORY POINTS

Story points are a relative unit of measure used to estimate the size and complexity of user stories. They help teams understand effort, uncertainty, and capacity without relying on time based estimates.

VELOCITY

Velocity is a measure of how much work a team completes in a Sprint based on story points. It is used to forecast future delivery and should never be used as a performance metric.

DESIGN & UX COLLABORATION

WIREFRAME

A wireframe is a low detail visual outline of a screen or interface. It shows layout, structure, and key elements without focusing on colors, fonts, or polish. Wireframes help teams align early on what will be built before development begins.

PROTOTYPE

A prototype is an interactive or higher detail model of a product that simulates how users will experience it. Prototypes are used to gather feedback, test assumptions, and refine designs before and during development.

UX

UX, or User Experience, focuses on how a user feels when interacting with a product. It includes usability, accessibility, clarity, and overall satisfaction.

UI

UI, or User Interface, refers to the visual elements users interact with, such as buttons, screens, layouts, and typography. UI supports UX by making interactions clear and intuitive.

JUST IN TIME DESIGN

Just in Time Design is the practice of designing only what is needed for the current or upcoming Sprint. This approach avoids over designing and allows the team to adapt designs as feedback is gathered.

DESIGN HANDOFF

Design handoff is the process of sharing design work with the development and QA team. In Agile teams, handoff is collaborative and continuous rather than a one time transfer.

DESIGN SPIKE

A design spike is a short time boxed effort used to explore uncertain design or technical questions. It helps reduce risk by allowing the team to learn before committing to full implementation.

USABILITY

Usability refers to how easy and efficient a product is for users to accomplish their goals. High usability reduces confusion, errors, and frustration.

USER FLOW

A user flow maps the steps a user takes to complete a task within a product. It helps teams understand how users move through screens and make decisions.

DEVELOPMENT & ENGINEERING PRACTICES

CONTINUOUS INTEGRATION

Continuous Integration is the practice of frequently merging code changes into a shared code base. Each change is automatically built and tested to catch issues early and reduce integration risk.

CONTINUOUS DEPLOYMENT

Continuous Deployment is the practice of automatically deploying validated code changes to an environment after passing tests. It allows teams to release updates quickly and reliably.

BUILD PIPELINE

A build pipeline is the automated sequence of steps that compile, test, package, and prepare code for deployment. It helps ensure consistency and quality across environments.

BRANCHING STRATEGY

A branching strategy defines how developers manage code changes in version control. It helps teams work in parallel while protecting the stability of the main code base.

ENVIRONMENT

An environment is a version of the system used for a specific purpose in the delivery process. Common environments include development, quality assurance, user acceptance testing, and production.

TECHNICAL SPIKE

A technical spike is a short, focused effort used to explore technical uncertainty or risk. It allows the team to learn and make informed decisions before committing to implementation.

CODE REPOSITORY

A code repository is a centralized location where source code is stored and managed. It allows teams to track changes, collaborate, and maintain version history.

AUTOMATED BUILD

An automated build is the process of compiling and packaging code without manual intervention. Automated builds reduce errors and provide fast feedback on code quality.

EXTREME PROGRAMMING

Extreme Programming, often called XP, is an Agile engineering approach that emphasizes frequent feedback, high quality code, and close collaboration. It promotes practices like pair programming, test driven development, and continuous improvement to reduce risk and improve software quality.

CODE REVIEW

Code review is the practice of having one or more team members examine code changes before they are merged. It helps catch defects, improve code quality, and spread knowledge across the team.

PAIR PROGRAMMING

Pair programming is a practice where two developers work together on the same code at the same time. One focuses on writing the code while the other reviews and thinks ahead, improving quality and shared understanding.

CODE REFACTOR

Code refactor is the process of improving the internal structure of code without changing its external behavior. Refactoring makes code easier to understand, maintain, and extend over time.

TEST DRIVEN DEVELOPMENT

Test Driven Development is a practice where tests are written before the code itself. Developers write a failing test, write just enough code to pass it, and then improve the design through refactoring.

BEHAVIOR DRIVEN DEVELOPMENT

Behavior Driven Development is an extension of Test Driven Development that focuses on defining system behavior in plain language. It uses scenarios and examples to describe how the system should behave from the user's perspective.

TESTING & QUALITY ASSURANCE

QUALITY STRATEGY

A quality strategy is the agreed approach a team uses to ensure the product meets quality standards. It defines how and when testing occurs, what tools are used, and how quality is shared across the team.

TEST CASE

A test case is a set of steps, conditions, and expected results used to verify that a specific piece of functionality works as intended. Test cases are often derived directly from acceptance criteria.

UNIT TESTING

Unit testing verifies that individual pieces of code function correctly in isolation. These tests are typically written and maintained by developers and run frequently.

INTEGRATION TESTING

Integration testing checks how different components of the system work together. It helps identify issues that occur when code modules interact.

REGRESSION TESTING

Regression testing ensures that new changes do not break existing functionality. It protects previously working features as the product evolves.

TEST AUTOMATION

Test automation is the use of tools to execute tests automatically rather than manually. Automated tests provide fast feedback and support frequent delivery.

BUG

A bug is a defect or error that causes the system to behave in an unintended way. Bugs are tracked, prioritized, and addressed through the backlog.

DEFECT

A defect is any deviation from expected behavior or quality standards. While often used interchangeably with bug, defect may include usability or performance issues in addition to code errors.

TRACEABILITY

Traceability is the ability to link requirements, acceptance criteria, test cases, and delivered functionality. It helps teams understand coverage and impact when changes occur.

ACCEPTANCE TESTING

Acceptance testing verifies that a user story or feature meets its defined acceptance criteria. It confirms that the work delivers the intended value.

USER ACCEPTANCE TESTING

User Acceptance Testing, often called UAT, is the validation of the product by business users or stakeholders. It ensures the solution meets real world needs before release.

DELIVERY, RELEASE & METRICS

RELEASE

A release is the act of making an Increment available to users. Releases may happen at the end of a Sprint or at any time when the product is ready.

DEPLOYMENT

Deployment is the technical process of moving code to an environment such as testing or production. Deployment does not always mean a release to users.

PRODUCTION

Production is the live environment where real users interact with the product. Changes in production have real impact and require strong quality and release practices.

ROLLBACK STRATEGY

A rollback strategy is a planned approach for safely reversing a deployment if issues occur. It reduces risk by ensuring the team can recover quickly.

MONITORING

Monitoring is the practice of observing system behavior after deployment. It includes tracking performance, errors, and usage to identify issues early.

TELEMETRY

Telemetry is the data collected from the running system about performance, reliability, and user behavior. It helps teams understand how the product performs in real conditions.

BURN DOWN CHART

A burn down chart shows how much work remains in a Sprint over time. It helps teams track progress and identify potential delivery risks early.

BURN UP CHART

A burn up chart shows how much work has been completed over time and how the total scope may change. It provides visibility into progress and scope growth.

CUMULATIVE FLOW DIAGRAM (CFD)

A Cumulative Flow Diagram visualizes how work items move through different stages of the workflow over time. It helps teams identify bottlenecks, work in progress limits, and overall flow health.

ESCAPED DEFECTS

Escaped defects are issues that reach users or production environments without being caught during development or testing. They indicate gaps in quality practices or feedback loops.

LEAD TIME

Lead time measures the total time from when a work item is requested to when it is delivered. It reflects how long customers wait to receive value.

CYCLE TIME

Cycle time measures how long it takes to complete work once it has started. It helps teams understand execution efficiency and identify delays within the workflow.

WORK ITEM AGE

Work item age measures how long a piece of work has been in progress. Rising work item age can signal blockers, bottlenecks, or over commitment.

TEAM MORALE

Team morale reflects the overall health, engagement, and emotional state of the team. High morale supports collaboration, quality, and sustainable delivery, while low morale often signals deeper system issues.

REAL TIME ADJUSTMENTS

Real time adjustments are changes the team makes during a Sprint based on new information. These adjustments help teams stay aligned with the Sprint Goal while responding to reality.

TEAM DYNAMICS & CONTINUOUS IMPROVEMENT

WORKING AGREEMENTS

Working agreements are shared guidelines the team creates to define how they communicate, collaborate, and make decisions. They help set expectations and reduce friction in day to day work.

PSYCHOLOGICAL SAFETY

Psychological safety is the shared belief that team members can speak up, ask questions, and raise concerns without fear of blame or punishment. It is essential for learning, quality, and continuous improvement.

BLOCKER

A blocker is any issue that prevents progress toward the Sprint Goal. Blockers are raised early so the team can resolve them quickly and avoid hidden delays.

RISK

Risk is the possibility that something may negatively impact delivery, quality, or value. Agile teams surface risks early so they can be addressed before becoming problems.

RISK RADAR

A risk radar is a shared way of identifying, tracking, and reviewing risks throughout delivery. It helps teams stay aware of uncertainty and respond proactively.

STAKEHOLDER

A stakeholder is any individual or group with an interest in the product or its outcomes. Stakeholders provide input, feedback, and support throughout the product lifecycle.

FEEDBACK LOOP

A feedback loop is the cycle of building, reviewing results, and adjusting based on what is learned. Fast feedback loops help teams improve quality and make better decisions.

CONTINUOUS IMPROVEMENT

Continuous improvement is the ongoing practice of reflecting on work and making small, meaningful changes over time. It is driven by learning, not perfection.

RETROSPECTIVE TECHNIQUES

Retrospective techniques are structured methods used during Sprint Retrospectives to help teams reflect and improve. Examples include Start Stop Continue, 4Ls, and Mad Sad Glad.

Get the
Scrum Quest
Answer Key
Scan the QR code to:
Download the official Scrum Quest Answer Key
Join the waitlist for upcoming resources
Get exclusive updates as Team Vertex's project documents, process mapping, backlog, wireframes, and prototypes are released
Start your next Scrum Quest — scan the code and get the Answer Key!

Made in United States
Orlando, FL
04 November 2026

90201517R00195

33
39
44
49
54
59

65
70
75
80
85

Andante
cresc.

Presto

207
f
211
p
215
f
219
p
223
f
p
229
f
233
p

Capriccio

Allegro assai

*) besser *ais*'? / better *a*'♯? / meilleur *la*♯'?

Allegro
Presto

Nachwort

Im 18. und 19. Jahrhundert war es weithin üblich, Werke, die ursprünglich für eine andere Besetzung bestimmt waren, in Klavierauszügen im Druck zu veröffentlichen. Die Klavierbearbeitungen dienten einerseits der Darbietung in privaten Zirkeln, andererseits aber auch dem Studium, solange Partiturausgaben selbst von Kammermusikwerken die Ausnahme darstellten. Bei vielen dieser Klavierarrangements sind die Namen der Bearbeiter nicht bekannt geworden. Wir wissen aber, dass Mozart in einigen Fällen selbst Klavierfassungen erstellt hat. Bei der *Entführung aus dem Serail* KV 384 kam der Verlag Schott in Mainz Mozart mit einem Klavierauszug zuvor. Mozarts eigener Klavierauszug, der bereits begonnen und über den Wiener Verleger Christoph Torricella angekündigt war, ist nie erschienen – mit Ausnahme der Ouvertüre.

Was für eine Offenbarung ist doch diese hervorragende Bearbeitung im Vergleich zu dem, was man in vielen Klavierauszügen dieser Oper findet! Das Achtel-Tremolo des Orchesters reduziert sich idiomatisch auf wiederholte Viertel, die ungestüme Bewegung und das Flair des Originals sind in einen vollkommen klaviergerechten Satz übertragen. Der Klang von „türkischen" Perkussionsinstrumenten und der Piccoloflöte bleibt im Ohr. Wie das Original, so endet auch diese Klavierbearbeitung mit einem Halbschluss, was bei einer eigenständigen Aufführung seltsam klingt. Hier könnte z.B. das *Capriccio*, das als viertes Präludium zu den *Vier Praeambula* KV 395 gehört, zur Abrundung eingesetzt werden. Es ist daher in dieser Einzelausgabe im Anschluss an die Ouvertüre abgedruckt.

Die Originalausgabe der Klavierfassung (Torricella, Wien 1785) ist weitgehend fehlerfrei, nur die Bögen sind häufig zu kurz, sie wurden durchweg an die Bögen in der Orchesterfassung der Ouvertüre angeglichen. Die wenigen Zusätze des Herausgebers sind im Notentext durch eckige Klammern kenntlich gemacht.

Ulrich Leisinger / Robert D. Levin

Postface

In the 18th and 19th centuries it was completely normal to publish in piano reduction works which were originally intended for other instrumentations. These piano arrangements provided on the one hand for performances in private circles, and on the other for study of the works, to such an extent that score editions even of chamber music were the exception rather than the rule. For many of these piano arrangements, the names of the arrangers are not known. We do know, however, that in a few cases Mozart prepared the piano version himself. For *Die Entführung aus dem Serail* ('The Abduction from the Seraglio') K. 384 the publishing house Schott in Mainz came out with a piano reduction before Mozart could do so. His own piano reduction, which he had already begun and had announced through the Viennese publisher Christoph Torricella, never appeared – with the exception of the overture.

What a revelation this superb arrangement is compared with what one sees in vocal scores of the opera. The quaver (eighth note) tremolo of the orchestra is reduced to idiomatic repeated crotchets (quarter notes); the brash color and flair of the original is preserved within a completely idiomatic keyboard texture. The sonority of the 'Turkish' percussion instruments and the piccolo should be kept in mind. Like the original, the overture ends on a half-cadence, making stand-alone performance awkward. Here, the *Capriccio* that forms the fourth of the 'Four Praeambula' K. 395 might come in handy. It is therefore, in this single edition, appended to the overture.

The original edition of the piano reduction (Torricella, Vienna, 1785) contains very few errors. Only the slurs are frequently too short. They have been adjusted throughout according to the orchestra version of the overture. The few editorial additions in the musical text are identified with square brackets.

Ulrich Leisinger / Robert D. Levin

Postface

Aux XVIIIe et XIXe siècles, il était d'usage de publier et d'imprimer, dans des adaptations pour piano, des œuvres initialement destinées à une autre distribution. Les adaptations pour piano étaient d'une part jouées dans des cercles privés, d'autre part, elles servaient également aux études, tant les partitions éditées, même d'œuvres de musique de chambre, étaient rares. Nombre de ces arrangements pour piano restent anonymes. Toutefois, nous savons que Mozart a, dans certains cas, procédé lui-même à l'adaptation d'une version pour piano. Pour « L'Enlèvement au sérail » KV 384, la maison d'édition Schott de Mainz a devancé Mozart avec la publication d'un extrait pour piano. L'extrait pour piano commencé par Mozart et annoncé par l'éditeur viennois Christoph Torricella, n'est jamais paru – à l'exception de l'ouverture.

Quelle révélation que ce superbe arrangement, comparé à ce que l'on trouve dans nombre de réductions pour piano de cet opéra ! Le trémolo de croches de l'orchestre se réduit idiomatiquement à des noires répétées, le mouvement impétueux et l'atmosphère de l'original sont traduits dans une composition pianistique absolument digne de l'instrument. On garde à l'oreille la sonorité des percussions « à la turque » et de la flûte piccolo. Comme l'original, cet arrangement pour le piano s'achève lui aussi sur une demi-conclusion, ce qui paraît étrange lors d'une exécution autonome. Ici, le *Capriccio* qui fait partie des quatre *Praeambula* KV 395 en tant que prélude peut être utilisé pour arrondir le tout. De ce fait, le *Capriccio* est imprimé à la suite de l'Ouverture dans la présente édition.

L'édition originale de la version pour piano (Torricella, Vienne 1785) est en majeure partie exempte de fautes, seuls les arcs de liaison sont souvent trop courts et ont été sans exception alignés sur la version orchestrale de l'Ouverture. Les rares compléments de l'éditeur dans la partition sont indiqués par des crochets.

Ulrich Leisinger / Robert D. Levin
(Traduction Micheline de Doncker / Geneviève Geffray)

Gedruckt mit 100% Ökostrom Pl. 7/ 2012